AF703027
WITHDRAWN

AGRICULTURAL RECORDS IN BRITAIN

A.D. 220–1977

Agricultural Records in Britain

A.D. 220–1977

J.M. STRATTON

and

JACK HOUGHTON BROWN

edited by

RALPH WHITLOCK

ARCHON BOOKS
Hamden, Connecticut

S
455
.S776
1979

First published in Great Britain 1969 as
Agricultural Records
Reprinted 1970
Second edition 1978
John Baker (Publishers) Limited

Second edition first published in the USA 1979
under the title
Agricultural Records in Britain A.D. *220–1977*
as an Archon Book, an imprint of
The Shoe String Press, Inc.
995 Sherman Avenue, Hamden, Connecticut 06514

© 1969 John Stratton; 1978 Ralph Whitlock

Library of Congress Cataloging in Publication Data

Stratton, John M.
Agricultural records in Britain, A.D. 220–1977.

Based on Records of the seasons, prices of agricultural
produce, and phenomena observed in the British
Isles, by Thomas H. Baker, first published in 1883.
First published in 1969 under title:
Agricultural records A.D. 220–1968.
1. Agriculture – Great Britain – History.
2. Crops and climate – Great Britain – History.
3. Great Britain – Climate.
4. Agricultural prices – Great Britain – History.
I. Brown, Jack Houghton, joint author.
II. Whitlock, Ralph.
III. Baker, Thomas Henry, 1833–1914.
Records of the seasons, prices of agricultural
produce, and phenomena observed in the British Isles.
IV. Title.
S455.S776 1979 630'.941 79-9895
ISBN 0 208 01818 2

Printed in Great Britain

Introduction

IN 1883 Mr Thomas H. Baker, who was farming on Mere Down, in south-western Wiltshire, had published a book entitled *Records of the Seasons, Prices of Agricultural Produce, and Phenomena Observed in the British Isles.* It dealt year by year with the weather, crop yields, agricultural prices and other matters of prime importance and interest to farmers, delving back to the very earliest times. In 1912, after his retirement, he prepared a revised edition, bringing the record up-to-date.

The present volume, built on Baker's work and prepared for Mr J.M. Stratton, who has farmed all his life land very near to Baker's own farm, takes the record up to the present day. It also attempts a broader assessment of the national pattern than Baker was able to make, his sources being mostly local.

These records are the bricks of which farming history is made—the unspectacular, prosaic, day-to-day facts with which every farmer is familiar. 1789, for instance, is a date which every schoolboy remembers as the outbreak of the French Revolution, but what were ordinary country folk doing in England while these dramatic events were shaping? They were, as the records show, coping with a late winter, a backward spring, a rather poor harvest and a serious outbreak of sheep-rot. The year of Waterloo was a year of drought but of abundant harvests, which were a contributory cause of a rapid fall in corn prices and consequent agricultural distress. And do you remember the rainy harvests of the early 1940s, when we were struggling to feed the nation in defiance of Hitler's threats?

Here is the unobtrusive background of history, a record of commonplace events which were far more important in the lives of ordinary country folk than were all the battles, treaties and conferences.

The Weather and the Crops

To what extent does the weather really affect our crops? Farmers have always maintained that the unpredictable British weather is their chief hazard. At the backs of their minds is a vision of the ideal climate, featuring moderate frost in winter; a dry February and March for sowing; plenty of showers and sunshine in April and May; a fine spell for haymaking in June and early July; then another instalment of refreshing showers before harvest; warm sunshine with no rain in August and early September; and an open autumn to enable them to get ahead with the ploughing and autumn sowing. When the weather strays from this ideal pattern, or even reverses it, which is pretty frequently, they consider themselves victimised. Drought in spring, thunderstorms at haymaking and rain in harvest will, they lament, ruin them.

What they say seems so logical that it is seldom questioned, but an examination of weather and harvest records does not altogether confirm it. In general, what farmers regard as unpropitious weather plays a far smaller part in crop yields than they like to think.

In the records of the past fifty years or so, let us look at some harvests which gave outstanding results, either good or bad.

In 1911, when the average wheat crop could be taken at about 16 cwt per acre and the average barley crop at about 14.5 cwt, wheat averaged 18.4 cwt per acre for the whole of England and Wales, and barley 15.5 cwt. The weather pattern does not conform to the ideal. 1911 was, in fact, a year of drought. April and early May, when rain is providential, were mainly dry, with cold northerly winds. The desired showery spell of July came a month early, interfering with haymaking. July, and August and September (to the 13th) were months of almost absolute drought and great heat, the temperature reaching 100°F at Greenwich on August 9th. The hay crop was light, and roots were badly affected by drought, but potatoes produced an average yield.

1921 was another very dry year, though the drought came earlier. April, May, June and July were the dry months. In England and Wales less than half the normal rainfall occured for the six months February to July. In this year the spring-sown cereals were certainly affected, all giving a less than average yield; and the same was true of roots and hay. On the other hand, autumn-sown wheat produced an average yield of 19.8 cwt per acre, which was an all-time record to that date.

1938 had its weather in exactly the reverse order to what farmers think they require. February, March and April were exceptionally dry, April being the driest since 1727. July was cool, dull and stormy, and August was distinguished by a remarkable series of thunderstorms in the first twelve days, followed by cooler, showery weather. Yet all crops yielded above average. Wheat hit the all-time high (to that date) of 20.3 cwt per acre, as the average yield for England and Wales; barley averaged 18.1 cwt per acre, which was also a record to that date; oats averaged 16.4 cwt per acre (equal to the previous highest record), and potatoes 7.3 tons per acre (unsurpassed to that date).

Going back farther, 1887 was one of the driest years of the nineteenth century. Streams dried up, villages had to be supplied with water from carts, and the level of Lake Derwentwater fell lower than had ever previously been recorded. Spring was late and backward, and August very hot. Yet wheat gave a bumper crop, and barley was up to average. There was also a good hay harvest.

The eighteenth century was noted for a long series of exceptionally dry years, beginning with 1714. That was one of the driest years ever recorded, the rainfall at Upminster in Essex being only 11.2 inches for the whole year, which is less than half the average for even that dry district. Spring was cold and dry, and in summer and autumn the drought was combined with heat. Yields of barley and oats were rather poor, but wheat was very good.

In 1741 a hot, dry summer followed a hard winter. Drought was almost absolute from January to the beginning of June; then, after a brief wet spell, it returned and lasted till the end of August. Yet the harvest was abundant. Beef sold for a penny a pound, wheat for sixpence a stone, and farmers complained that they were being ruined by plenty.

1742 and 1743 were similarly dry years, but both produced excellent harvests.

Now let us turn to excessively wet years. 1852 had a very dry spring, followed by exceptionally heavy and persistent rains in August, September and October, accompanied by high temperatures. There was consequently much damage to the crops, but the harvest was heavy, even although the quality was poor.

1872 was different in that spring was cold and wet, June and July distinguished by thunderstorms, August and September mainly fine, and the rest of the autumn excessively wet. This should have made a better pattern than 1852, yet harvest was, on the whole, poor. Cereal crops were much damaged by the July storms.

1879 was one of the wet years of the nineteenth century. Spring was very backward, with cold weather, including snow, continuing well into May. June, July and August brought abnormal rains, but September was finer and enabled most of the harvest to be gathered. This year the rains coincided with a poor harvest.

1903, on the other hand, was a year of unusual rainfall, with excessive rain both early in the year and then, following a dry spring, in summer and autumn; yet it gave crops above average.

1924 was one of the dreariest years, with cool, dull weather as well as abnormal rains right through the summer. Rainfall in general ranged from 130% to 180% of average. Yet nearly all crops gave about average yields, except wheat, which was rather above average.

We will look at a few years when harvests were below average.

1920 had spring crops a little below average and winter wheat very much so. The pattern for the first three months of the year was very near the ideal; but the summer months were dismal. They were not only wet but cloudy and cool, with numerous frosts in June and August.

1930 had a late spring, a sunny and warm June, and then cool, damp weather till after harvest, though with a few fine spells. The harvest was deficient for nearly all crops.

1937, another wet year, was also distinguished by lack of sun. Crops were poor, especially barley.

1947, the last year of really poor harvests, was a year of extremes. Late January, February and the first half of March saw a period of intense cold, with heavy snow. April and May were unusually wet. The rest of the summer brought almost ideal conditions, June and July being warm with thunderstorms, while August and September brought heat and droughts. Yet harvests were deficient all round, crop yields in general being back to the levels of the 1920s.

What can be gathered from all these apparent contradictions?

First, they provide a vindication of the old farming axiom that 'there is never a year in which every crop succeeds or one in which every crop fails'.

Examined more closely, the records indicate that the reason for this is that the pattern of weather which suits winter wheat may be adverse to spring corn, and vice versa.

We may go further and postulate that the key factor is the weather at sowing-time or just after. Once a crop has germinated and tillered, the subsequent weather does not greatly matter.

Winter wheat usually has its fate decided by the weather in autumn.

Extreme cold in winter can damage it, though not, as a rule, if accompanied by snow. In spring it revels in a timely spell of rain, though good crops have been produced after a spring drought.

Spring crops are, of course, dependent on the spring weather, though, again, once they have become established they can endure extremes. Oats, on the whole, are less susceptible to weather vagaries than is barley, perhaps because they are indigenous to Britain.

Potatoes and roots naturally flourish in damp summer more than in dry. Hay is of all crops the most vulnerable to rain, and the records lend emphasis to the view that no-one in his right senses would attempt to make hay out-of-doors in our climate in three years out of five.

Apart from hay, few crops are affected, in yield, by rain after about the middle of June, though the subsequent weather does, of course, determine the quality and the amount of waste. A dull summer is as detrimental as a wet one.

It must be stressed, however, that we are dealing with averages. The weather pattern varies considerably from place to place in Britain, and individual farmers may be able to show that their own crop records are much more closely linked with the local weather than the national figures might indicate.

Finally, since 1947 the weather seems to have made little difference to the upward climb of the graph of crop yields. After a 1947 average of 15.2 cwt per acre of wheat, the average shot up to 20.7 cwt in 1948, and thereafter progress has been pretty steady. In 1950 the average was 21 cwt, in 1960 it had risen to 28.4 cwt., and in 1965, in spite of a wet summer, it stood at 32.4 cwt. Barley and oats show similar improvement, and the potato crop average has now risen to 10.6 tons per acre. It seems that improved plant varieties, improved fertilizers, adequate harvesting equipment and efficient grain-driers are at last giving us an ascendancy over the weather.

R.W.

Preface to the Second Edition

The steady and consistent demand for *Agricultural Records* has prompted the preparation of this new edition, which takes the records of agricultural prices, seasons and other phenomena to the end of the year 1977. The task of updating the chronicle has again been undertaken by Ralph Whitlock, at the instigation and with the help of Colonel Jack Houghton Brown, nephew of the late Mr. John Stratton, and who now farms much of the land on Mere Down that provided a livelihood in the late nineteenth century for Thomas Baker, the compiler of the original Records.

A new feature is the table of agricultural prices going back, though with many gaps, to 1257. Obviously the figures given for the earlier centuries can be only approximate, the prices of most commodities naturally varying from market to market, but they do nevertheless reflect the quality of the harvest and the state of agriculture (then the nation's prime industry), thus providing an interesting commentary on national events. The annual records of the 1970s illustrate the impact of the decimalisation of the coinage, followed by metrication and the general adjustment to new conditions imposed by the entry of Britain into the European Economic Community.

NOTES

A tod of wool weighs 28 lb.

A quarter of wheat is 4½ cwt.

The prices quoted for grain in the later years of these records does not include subsidy payments.

Agricultural Records

A.D.

220 A great frost in England is said to have lasted 5 months.

230 The Thames is said to have been frozen over in London for 6 weeks.

245 Many thousands of acres in Lincolnshire were flooded by the sea.

272 A year of famine in England.

291 Many English rivers said to be frozen for 6 weeks.

300 A great famine in Scotland.

306 A year of famine in Scotland.

310 A great famine in England.

329 Many rivers in Britain said to have been frozen for 6 weeks.

353 A great flood in Cheshire in which 5,000 people and many cattle are said to have perished.

359 A year of plague in Britain, with heavy mortality. Very severe frost for 14 weeks in Scotland.

430 Disastrous year for plague in England.

447 A year of famine in England, accompanied by plague.

508 All the rivers of Britain were said to have been frozen for more than 2 months.

545 A winter of intense cold.

554 Another very severe winter with much snow.

558 A year of plague.

664 A great drought, accompanied by plague.

665 Another year of severe plague.

678 Around this time there was a three years drought in Southern England.

686 A summer of plague in England.

695 The river Thames was said to have been frozen over for 6 weeks.

738 Great flood in the Glasgow region.

739 A year of famine.

759 A winter of extraordinary frost from October 1st, 759 to February 26th, 760.

761 Another exceptionally cold winter.

763 A summer of drought and great heat.

764 Another winter of extreme cold with heavy snowfalls, followed by a late spring.

772 A year of plague in England.

793 A year of famine.

795 Another year of famine.

797 Drought and famine in England.

799 A year of storms in which many ships were lost around the coasts of Britain.

800 On Christmas Eve (old style) a great gale from the south-west did immense damage to property, destroying innumerable houses and tearing up many trees. This was followed

later in the year by extensive sea floods and by epidemics among cattle.

810 Very severe cattle plague in England.

823 A year of famine, causing great mortality.

827 A winter of great frosts which lasted for 9 weeks.

836 Extensive floods in the border country between England and Scotland, with the river Tweed overflowing its banks.

874 The beginning of a great frost in Scotland, which lasted from early November, 874 till the end of April, 875. This frost was followed, when the thaw came, by great floods.

877 This is the year in which 120 Danish ships are reported to have been lost in a great storm at Swanwick (Swanage).

892 A severe late frost which is reported to have killed many vines and caused mortality among cattle.

896 Heavy mortality through disease among men and cattle.

897 A year of plague among cattle and human beings.

898 Another year of late frosts.

908 Many English rivers frozen for about 2 months.

923 The river Thames was frozen for 13 weeks in London.

926 In this year by the law of King Athelstan a sheep was valued at 1s.; an ox at six times the value of a sheep; a cow at four times the value of a sheep. A horse was valued at 30s., a mare at 20s., and a man at £3. The value of a fleece was two-fifths of the value of the sheep.

944 A great storm destroyed about 1,500 houses in London. This storm also did much damage in other parts of England.

954 A year of famine throughout Britain, which lasted, off and on, for 4 years.

962 A year of plague in Britain.

970 A year of storm, with excessive amount of cloudy weather.

971 Another very cloudy year, the sun being hidden most of the time for 6 months.

974 A great earthquake in England.

975 A year of famine in England.

976 Another year of terrible famine.

981 The early part of the year was distinguished by a very severe winter which lasted late into spring. Famine and plague followed later.

986 A year of plague, with great mortality, among cattle.

987 Cattle mortality continued, and an epidemic also carried off great numbers of people.

991 The year began with extremely severe and long winter. Many crops failed, and in autumn famine and plague occurred.

993 Very hot, dry summer.

994 Another summer of drought and heat, with consequent damage to corn and fruit.

998 The river Thames is said to have been frozen over for 5 weeks.

1000 A general failure of crops and mortality among cattle.

1001 A year of plague.

1005 A very great famine throughout England.

1007 Another great famine, consequent partly on an unfavourable season but also on the disturbed state of the country.

1008 Very great storms occurred.

1009 Another tremendous storm which caused great loss to shipping.

1013 Very wet year.

1014 On September 28th a great sea flood inundated great areas of the country.

1015 Another very severe sea flood with the spring tides. Much mortality.

1016 A year of famine.

1022 A very hot, dry summer.

1039 A very severe storm occurred, month unrecorded.

1040 Sea flood caused great inundations in summer. On midsummer day severe frost damaged the crops and destroyed much fruit. In consequence there was a great shortage of food in autumn.

1041 A year of storm and rain, with much consequent disease. This began a series of famine years which lasted till 1066.

1043 Famine increased, with corn reaching excessive prices.

1044 Another severe famine year.

1045 The year began with excessively severe winter with much snow, causing heavy mortality among men and livestock.

1046 Another famine year with outbreaks of plague.

1047 A great snowfall occurred, beginning early in January and continuing, off and on, until March 17th. A stormy summer, with many thunderstorms. The famine continued in England.

1048 The year began with very severe winter. Another year of shortages, with heavy mortality among men and animals. A severe earthquake was felt in the Midlands in May.

1049 Heavy mortality among men and animals continued, caused by famine and plague.

1051 Famine continued.

1052 A stormy year.

1053 A very severe gale in winter did much damage to property.

1054 Great mortality among cattle, following a severe winter.

1063 The Thames was said to have been frozen for 14 weeks.

1069 Extremely hard winter began the year. A great famine occurred throughout much of England but especially in the North, where there was heavy mortality.

1070 Severe famine continued.

1071 Severe famine continued.

1072 Very hard winter.

1076 Severe frost which began on November 1st lasted until the middle of April 1077.

1078 Very dry summer.

1082 The year of famine.

1083 A very heavy rainfall in the autumn, at the end of October and in early November, caused great floods and much loss of life.

1085 Very cold weather.

1086 A year of storm and of pestilence among cattle.

1087 A year of great famine, following many severe storms. There was also an epidemic of plague which carried off large numbers of people, and also cattle.

1088 A very late harvest with some crops not ripe till the end of November. Consequent shortages.

1089 Another very late harvest, extending into November.

1090 A very late harvest and a great scarcity of fruit. In November a severe storm caused flooding which damaged London Bridge.

1091 Several excessively severe storms occurred this autumn. On October 5th about 500 houses were destroyed in London. On the same date the steeple of a church at Winchcomb, in Gloucestershire, was struck by lightning and destroyed. On October 17th another storm did much damage in London, and also in Salisbury (Old Sarum), where the top of the steeple was thrown down. On November 6th London Bridge was swept away by floods.

1092 The high spring tides caused great flooding in England and Scotland. It was at this time that the estates of Earl Goodwin, on the Kent coast, were inundated to form the Goodwin Sands. Severe frost followed later in spring, later in the year more wet weather occurred causing renewed flooding.

1093 Another year of severe floods followed by great frosts. Much damage was caused.

1094 Great scarcity of food in England, with some consequent famine and plague. Heavy mortality.

1095 A year of storm and tempest, with adverse affects on the crops.

1096 A year of great famine.

1097 Another year of storm and rain.

1098 Another very wet year with much damage to crops. The *Anglo-Saxon Chronicle* says that 'the great rains ceased not all the year.'

1099 A great inundation by the sea at high tide in November.

1100 Much damage caused by sea floods at high spring tides.

1103 Great gale of wind on August 10th did much damage to crops. Pestilence also attacked both men and animals.

1105 Another bad year for crops.

1109 A year of storms, with much thunder.

1110 Much damage to crops by heavy storms in summer.

1111 The year began with exceptionally long winter. This was followed by famine and plague causing terrible mortality among both animals and men.

1112 A very good year for crops came in with an abundant harvest, but plague caused heavy mortality.

1114 For some reason the Thames ran very low on October 15th, so that people could wade across it below London Bridge. Severe gales and wind also occurred in October. This was followed by a very severe winter, with great frosts.

1115 This year began with an exceptionally severe winter with snow and frost causing much mortality among cattle.

1116 The year began with a very severe winter. In summer excessive rains in July and August caused a failure of many crops. An exceptionally heavy thunderstorm occurred on October 31st.

1117 A very wet summer, with consequent harvest failures. A severe thunderstorm occurred on December 1st.

1119 A severe earthquake recorded in Gloucestershire and Worcestershire on September 29th.

1121 On Old Christmas Eve a great storm occurred.

1122 Severe storms were recorded on March 22nd and September 28th, with much damage to shipping. An earthquake occurred in the West Country on July 25th.

1124 A year of harvest failures with consequent high prices of corn.

1125 A year of famine. It began with intense cold in the winter, followed by excessive rain and floods during harvest. Prices for corn rose to excessive heights.

1126 Another year of great famine with corn rising to unprecedented prices.

1131 Severe epidemics among cattle and other domestic animals.

1133 Severe earthquake shock felt in the first week of August.

1134 A sudden destructive sea flood along the East Coast on October 1st.

1137 A year of famine, with very poor harvest and corn excessively dear.

1140 A destructive tornado hit Wellesbourne, Warwickshire, damaging many buildings and accompanied by a violent hailstorm.

1142 The Thames was frozen over in London on Old Christmas Day.

1147 Very hot summer, especially in August.

1149 Very heavy rain in harvest did much damage to crops. Very intense frost began on December 10th and continued until February 19th, 1150. The Thames was frozen over at London Bridge, supporting even loaded waggons.

1150 A year of famine.

1151 Another year of severe famine.

1152 Very wet summer caused poor harvests and much shortage of food.

1153 Another year of famine with much mortality, especially in Scotland.

1158 A dry year, in which the river Thames dried up so that it could be crossed at London on foot.

1171 An exceptionally severe thunderstorm throughout England on Christmas Eve (old reckoning).

1172 A sudden and severe thunderstorm on Christmas Day (old reckoning).

1174 A severe gale on December 31st.

1175 A year of famine and plague.

1176 Unusually heavy snow, followed by a period of hard frost characterised the early part of the year. On Midsummer Day severe hailstorm killed many sheep and young cattle.

1178 Severe snow-storm in early January, accompanied by much drifting was followed by floods when the thaw came.

1189 An exceptionally severe thunderstorm occurred on March 7th. This was a year of famine, occasioning many deaths.

1190 Severe shortage of grain following drought.

1193 A year of famine.

1194 The famine continued.

1195 Another year of famine but the harvest was spoiled by excessive rains and floods.

1196 Another year of famine accompanied by plague. Heavy mortality.

1198 A year of scarcity and famine in Scotland.

1199 A year of good harvests in Scotland, but heavy rain and much flooding in England.

1200 After an exceptionally cold winter, spring brought continuous rain and severe floods.

1201 Another exceptionally wet year when much hay and corn was spoiled or carried away by floods.

1202 The year began with an exceptionally cold winter. Holinshed records 'ale was frozen within houses and cellars and sold by weight. Such a great snow fell also therewith that beasts died in many places in great numbers.' During the summer frequent storms occurred, doing much damage to harvest.

1203 A year of heavy rains in London.

1205 A frost that began on January 15th and continued until March 22nd causing much delay in cultivations and spring sowing. In the following summer wheat was extremely dear, about ten times its normal price.

1207 On January 27th a tremendous storm, accompanied by snow caused much damage to property and killed many sheep and cattle. Subsequently the cold was intense.

1210 Floods in Scotland at Michaelmas.

1212 Dry summer. A great fire in London.

1214 Another dry summer, in which the Thames was so low in London that women and children could wade across it.

1222 A year of scarcity and famine. A great storm with thunder and lightning occurred on February 8th. The spring and summer proved unusually wet, and the sea tides were much higher than normal. Much of the harvest was ruined by the weather.

1223 A year of cattle plague.

1224 Severe storm which caused much damage on October 18th.

1228 A very wet summer with many thunderstorms. Much damage and delay to harvest.

1233 On March 23rd a tremendous thunderstorm heralded the beginning of a very wet spring and summer. Floods occurred in many places, and harvest was greatly damaged and delayed. The weather was also cold. A year of famine.

1235 An exceptionally wet spring. The Thames rose to unusual heights.

1236 Another exceptionally wet year. January, February and early March brought unusually heavy rainfall, consequently rivers rose and caused much flooding. In February it is said that the rain fell for eight days without ceasing. Autumn too, was very wet with much flooding and high tides along the coasts. There was much loss of cattle and also of human life.

1237 Another very wet spring, beginning with heavy rain around March 1st, but the year was not quite as wet as 1236.

1239 The first 4 months of the year were exceptionally wet.

1241 A year of pestilence.

1242 An exceptionally heavy storm which began on November 20th and continued for many days caused the rivers to rise to unusual heights, with much flooding.

1243 A stormy year, yet recorded as 'tolerably fertile and fruitful'.

1244 Another year of good harvests especially corn.

1246 A discouraging year. On April 25th a severe frost with snow caused much damage to fruit trees and crops in the fields. On July 20th a severe thunderstorm began a wet period which caused damage to the harvest.

1247 February 14th heralded a month or more of exceptionally heavy rain, with low temperatures. Much damage was done to fruit trees, but field crops subsequently yielded well.

1248 Another favourable year with an abundant harvest of corn. Fruit was also plentiful but in some districts a plague of worms and grubs caused much damage to fruit trees.

1249 Another fairly abundant harvest, although some damage was caused to it by heavy rains in June. Holinshed records that by a storm in June 'the corn in the field was so beaten to the ground that bread made thereof, after it was ripe, seemed as it had been made of bran'.

1250 Apparently a very stormy autumn.

1251 A year of famine and plague. A great storm on May 19th did much damage to property and cattle. At Michaelmas there were serious sea floods.

1252 A four months' drought which seriously affected crops. This drought began at the end of March and was at first accompanied by morning frosts and strong northerly winds which did much damage, especially to fruit trees. Holinshed records, 'the grass was so burned up in pastures and meadows that if a man took up some of it in his hands and rubbed the same ever so little, it straight fell to powder, and cattle were ready to starve for lack of meat, and because of the exceeding hot nights there was such abundance of fleas, flies, and gnats, that people were vexed and brought in case to be weary of their lives. And herewith chanced many diseases, as sweats, agues, and other. In the harvest time fell there great death and murrain amongst cattle, and specially in Norfolk, in the fens and other parts of the south. This infection was such that dogs and ravens feeding on the dead carrion swelled straightways and died, so that the people durst eat no beef. The cause of the death of cattle was thought to come hereof. After so great a drought, which had continued by all the space of the months of April, May, June, and July, when there followed good plenty of rain, the earth began to yield her increase most plenteously, and so the cattle, which before were hunger starved, fed now so greedily of this new grass that they died. Apple trees and pear trees began again to blossom after the time of yielding ripe fruit'.

1253 Another year of drought, at least in spring and early summer. In harvest a rainy spell intervened, causing floods that did much damage. At the end of harvest however, about Michaelmas, another drought occurred, so severe that many mill-streams dried up, so that it was impossible to grind corn. A great snowstorm occurred on December 13th, and this was followed by an exceptionally severe winter which lasted until the middle of March in the following year.

1254 The year began with an exceptionally cold spell which lasted till the middle of March. Plague also occurred among sheep, with heavy mortality. The spring was characterised by cold northerly winds which delayed the growth of crops. A very heavy hailstorm occurred on July 1st causing much damage. Disease called 'tongues' evil' was responsible for the death of many horses in late autumn.

1255 Severe gales, with heavy rain, which began on February 14th were followed by a very wet and unseasonable spring. Later the situation was aggravated by a drought throughout the whole of April.

1256 An exceptionally severe thunderstorm, with subsequent flooding, in the Eastern Counties on August 11th.

1258 A year of famine. Spring was unusually late, with northerly winds persisting until well into June. On June 24th a tremendous storm caused flooding and much damage to cornfields in the west of England, and also occasioned some loss of life. Owing to almost continuous rains, the harvest was very late, some of it not being gathered until November 1st. Many people in England died of famine this year.

1259 A harvest which promised to be not nearly as abundant as that of 1258, and yet yielded more, because of very favourable harvesting weather.

1260 This year started with a very mild period followed by a late and unseasonable spring. Corn crops were deficient in yield, but there was an abundance of fruit. A tremendous thunderstorm occurred on July 27th, with a very heavy fall of hail. Although this is said to have been a year of famine in England, corn on the other hand seems to have been reasonably cheap, wheat being 4s 3d per quarter as against 13s 5d in 1247, another famine year.

1262 The year began with a period of severe frost, afterwards it proved to be a year of abundance with plentiful harvest.

1263 A year of scarcity with a poor harvest.

1264 Another year of deficient harvest.

1265 Apparently a good harvest this year, as the price of corn fell.

1266 A moderately good harvest year. Apparently much rain with floods and high tides.

1268 The year began with a very hard winter, with the Thames frozen over at London. A very heavy thunderstorm in April was followed by a fortnight of heavy rain.

1269 Exceptionally severe rains in early February, with consequent flooding in London and elsewhere.

1270 Apparently not a very good harvest, as wheat prices rose.

1271 Wool $3\frac{3}{8}$d per lb.

Wheat prices again high, reaching 10s per quarter at one time.

1273 Corn prices had fallen to their former levels, evidently through fairly good harvests.

1275 Disease said to have been introduced from Spain attacked sheep.

1276 An exceptionally hot and dry summer, with a consequent scarcity of fodder.

1277 Wool 9d per lb.

Wool was very dear, probably because of the recent appearance of scab among sheep. Another hot and dry summer with consequent scarcity of fodder.

1278 Another year of drought.

1279 Wool 3d per lb.

An early harvest.

1280 Apparently a very good harvest.

1281 Wool 4d per lb.

An exceptionally cold spell with the very heavy snowfalls occurring in the first months of the year. The Thames was frozen over at London. Later in the year there was a severe drought.

1282 Evidently a year of scarcity with corn prices high. Wool 8s 9d per tod.

1283 Heavy mortality among sheep.

1284 A summer of drought. Wool 8s 1d per tod.

1286 Apparently a very good harvest. Wool 9s 3d per tod.

1287 An abundant harvest. Holinshed records 'was such plenty of grain that wheat was sold in some places for twenty pence a quarter, and in some places for sixteen pence, and peas for twelve pence a quarter.' Autumn brought a very wet period with consequent floods, especially along the east coast. In December some 500 people lost their lives in a sea flood that inundated Norfolk. This was the flood that swept over Holland, creating the Zuider Zee and claiming the lives of 50,000 people. Wool 8s 1d per tod.

1288 A very hot, dry summer. Scab became a serious disease among sheep. Wool 9s 8d per tod.

1289 A very wet summer, beginning with a heavy storm on July 9th. Consequent scarcity of food. Wool 9s 6d per tod.

1293 The price of wheat climbed to the unusual height of 10s 6d per quarter. A rainy harvest. On May 14th a heavy snowstorm occurred. Wool 8s 8d per tod.

1294 A year which promised abundant harvest, but the summer was so rainy that much of the corn was destroyed. Floods followed in many places in autumn. The price of wheat rose to 12s per quarter. Wool 6s 2½d per tod.

1299 A year of abundant harvest, but a bad year for fruit. Wool 8s 4½d per tod.

1300 A very dry summer, with some farmers unable to make hay at all. Despite this there was apparently a good harvest. Wool 8s 9d per tod.

1305 A year of drought and great heat. The hay crop failed through lack of rain and many animals died for want of grazing. A severe epidemic of smallpox and possible some other disease caused great mortality among human beings. A very severe spell of frost began on December 15th. Wool 8s 3½d per tod.

1306 Frost which began on December 15th, 1305 lasted until January 25th and caused much damage, the temperature being exceedingly low. After a mild spell, the winter returned and brought severe weather from February 13th till April 13th. Wool 8s 11d per tod.

1309 Period of severe frost was followed by a sudden thaw which caused much flooding. One such sudden flood occurred in Salisbury, when the Cathedral was flooded. This was the second of two rainy summers. Wool 8s 9d per tod.

1310 Another wet harvest following a severe winter. Wool 8s 1d per tod.

1311 Another cool, stormy harvest. Wool 7s 4d per tod.

1312 Yet another wet harvest. Wool 8s 7d per tod.

1313 A year of scarcity and famine following the period of wet deficient harvest. Many cattle have died from malnutrition. Wool 11s 2d per tod.

1315 This is called the first year of the great famine. It was another wet year with exceptional rain during July and August, although the crop which was ready for harvest then was not good. It is said that the crops in England were almost a total failure. The price of wheat rose to 26s 8d per quarter. There was heavy mortality among human beings and cattle, and the situation was aggravated by plague among cattle. Wool 9s 4d per tod.

1316 Another year of famine. Hume records, 'the kingdom of England was afflicted with a grievous famine during several years about this time. Perpetual rains and cold weather not only destroyed the harvest, but bred a mortality among the cattle, and raised every kind of food to an enormous price.

Wheat was sometimes sold at 90s a quarter.' Cattle were still afflicted by plague. Wool 9s 5d per tod.

1317 A deficient harvest, although much better than in the previous two years. Until harvest time the famine continued and great suffering occurred throughout England. Wool 8s 7d per tod.

1318 Apparently a better harvest for things were now returning to normal. Poor hay crop, and plague continued among cattle. Wool 9s 10d per tod.

1319 Another wet harvest and much mortality among cattle through 'murrain'. Wool 11s 9d per tod.

1320 Wool 9½d per lb.

Apparently another poor harvest. Disease called 'Ffarsine' caused much trouble among horses.

1321 Still a time of scarcity, with high prices and apparently another deficient harvest. Wool 10s 9d per tod.

1322 Evidently a much better harvest, for in the following year prices of corn fell considerably, wheat costing 5s a quarter in the following June. Wool 9s 8d per tod.

1324 A summer drought with scarcity of green fodder throughout England. However, the harvest seems to have been quite good. Wool 8s 3d per tod.

1325 Another year of drought, the summer being exceedingly hot and dry. However the harvest was apparently very good. Wool 10s per tod.

1326 Evidently another favourable harvest. Wool 10s 6d per tod.

1327 A cold, rainy harvest. Wool 10s 8d per tod.

1328 A poor hay harvest, but better weather for the corn harvest. Wool 8s 10d per tod.

1329 A bad year for fruit, and apparently the harvest was somewhat deficient. Wool 7s 4d per tod.

1330 A spring drought followed by heavy rain in summer and autumn. This rain began on July 16th and was so continuous that in some places harvest did not begin till Michaelmas. Much of it was not gathered until November. Wool 9s per tod.

1331 A mild spring and favourable summer, with comparatively little rain. An excellent year for cider. Wool 10s 6d per tod.

1332 An excellent harvest and general prosperity in the countryside. Wool 8s per tod.

1333 A year of drought. Much disease among flocks of sheep. A severe sea flood on the night of November 23rd along the east coast. Wool 5s 10d per tod.

1334 The year began with a prolonged frost, and the harvest appears to have been not very good. It is said that the yield of wheat about this time seldom reached two quarters per acre and was generally no more than one. Barley and oats also yielded about two quarters per acre on an average. Wool 7s 1d per tod.

1335 An excessively wet year. In consequence the corn harvest was a failure, and disease was rife among cattle and sheep. Wool 8s 4d per tod.

1337 A very favourable harvest year. Wool 5s 11d per tod.

1338 The harvest this year is said to have been the most abundant since 1287. Wool 6s 9d per tod.

1339 In the previous autumn an excessive rain hindered the sowing of winter corn. In December frost, which lasted for three months, began destroying most of the corn that had had been sown. A famine resulted in Scotland, most crops failing, but the harvest seems to have been not too bad in England. Wool 6s per tod.

1342 A famine in England and Scotland, following a poor harvest in 1341. Wool 6s 6d per tod.

1343 A very dry year, but with apparently fairly good harvest. Wool 8s 9d per tod.

1344 Another year of drought. Wool 7s 8d per tod.

1345 A year of unusual drought, rain being particularly lacking in spring. Scab very prevalent among sheep. Wool 7s 11d per tod.

1347 A bad harvest, and a very poor lamb crop. Beginning of the Black Death among human beings. Wool 7s 6½d per tod.

1348 An excessively wet summer, with almost continuous rain from mid-summer to Christmas. Serious floods followed, and much of the harvest was not gathered. The Black Death caused very heavy mortality. Wool 5s 8d per tod.

1349 The Black Death continued, and moved also into Scotland where it caused very heavy mortality. Wool 4s per tod.

1350 Cold and wet summer, with general scarcity of provisions and the plague continuing, at least until August. Wool 6s per tod.

1351 Another famine year with a cold, wet summer and high prices for corn.

1352 An abundant harvest. Wool 5s 6d per tod.

1353 A great drought from the end of March till the end of July. In consequence there was a great scarcity of grass and later the harvest was deficient. Some famine in England. Wool 8s per tod.

1354 Another year of drought, with a deficient harvest and no fruit at all. Wool 7s 2d per tod.

1355 A wet summer but not a very good harvest. Wool 6s 4d per tod.

1356 Another wet summer with flooding towards the end of harvest. Much damage to crops. Wool 5s per tod.

1357 Apparently a rather better harvest. Wool 7s 4d per tod.

1361 A year of drought. Wool 7s 10d per tod.

1362 A drought in spring and summer with deficient harvests resulting. However, other writers say that the harvest period was wet with much corn and hay spoiled. Possibly the drought occurred in early summer and the rainy weather later on. A tremendous storm blew up on January 15th and did immense damage, destroying many tall buildings and uprooting many trees.

1364 This year a very severe frost, which began on December 7th 1363 lasted until March 19th, greatly delaying spring cultivations and causing much damage to crops. However, fairly good harvest seems to have resulted.
Wool 9s per tod.

1366 Very wet period around mid-summer, causing much hay and some corn to be lost. A recurrence of plague or some other diseases caused many deaths.
Wool 10s per tod.

1368 Apparently a poor harvest, for grain prices rose to excessive heights in the following year. Wool 8s 10d per tod.

1369 A year of famine, with much mortality among men and animals. Wool 9s 4d per tod.

1370 Another very wet summer, with much corn spoiled at harvest time. However, in spite of this fair yields seem to have been gathered. Wool 9s 7d per tod.

1371 A dry summer. Wool 9s 8d per tod.

1376 A year of drought but with good harvest and corn much cheaper than in recent years. Wool 11s 1d per tod.

1377 Another drought in spring and summer, but apparently an abundant harvest. Wool 12s 7d per tod.

1379 The wheat harvest seems to have suffered but other corn yielded well. The fruit harvest failed. Wool 10s 11d per tod.

1380 An excellent harvest. Plague caused heavy mortality in Scotland. Wool 9s 4d per tod.

1383 An unseasonable cold spell during March and part of April, but the harvest seems to have been quite good. Wool 10s 3d per tod.

1384 A very severe winter. Wool 8s per tod.

1385 Apparently a good harvest year but heavy mortality among sheep. Wool 9s 2d per tod.

1386 A good harvest year. Wool 7s 3d per tod.

1387 Another good harvest year. Farmers were allowed reduction of rents on account of the low corn prices. Wool 7s 11d per tod.

1388 A rather poor harvest. Wool 7s 3d per tod.

1389 Apparently this also was a deficient harvest. Much mortality among sheep. Wool 7s 5d per tod.

1390 The prices of grain rose to high levels in the early part of the year, and near famine conditions prevailed, but afterwards a favourable harvest apparently corrected this state of affairs. Wool 7s 6d per tod.

1391 An excellent harvest. In the following spring wheat was said to be lower in price than for a hundred years. Wool 8s 2d per tod.

1393 Apparently a very good harvest, the summer was hot and dry, but the fine spell broke in thunder during September and was followed by considerable flooding. Wool 7s 8d per tod.

1394 Another hot, dry summer with apparently a good harvest. Wool 7s 10d per tod.

1395 A backward spring and evidently a cloudy and wet summer, with lower yields than in past years. A deficient fruit crop. Wool 7s 8d per tod.

1396 Another rainy summer, but with an average harvest. Wool 7s 2½d per tod.

1397 A good harvest, but no fruit. Wool 8s 11d per tod.

1398 A good harvest, but a very wet autumn. Again a deficient fruit crop. Wool 8s 3d per tod.

1399 Apparently a rather poor harvest. Wool 8s 2d per tod.

1400 Evidently another wet summer, with both hay and corn harvests suffering. Wool 8s 2d per tod.

1401 A wet harvest but apparently yields were quite good. Wool 7s 6¾d per tod.

1402 Another quite good harvest, though a wet period in mid-August interrupted harvesting. A bad fruit year. Wool 9s per tod.

1403 A good harvest. Wool 8s 11d per tod.

1404 An abundant harvest. Wool 8s 10¼d per tod.

1405 Another excellent harvest. The year is said to have been one of 'great plenty'. Wool 9s 1¼d per tod.

1406 Another good harvest. Wool 9s 4½d per tod.

1407 This year began with a prolonged period of frost and snow which lasted from December to March and caused much mortality among birds. The harvest was heavy but not of invariably good quality. Wool 11s 3½d per tod.

1408 A poor harvest, especially in eastern counties. Wool 9s 5¼d per tod.

1409 A wet summer, but apparently a reasonably good harvest. A bad fruit year. Wool 9s 3d per tod.

1410 Another good harvest but again no fruit. Wool 9s 3d per tod.

1411 A hot, dry summer with an abundant harvest. Again no fruit. Wool, 8s 3¾ per tod.

1412 Again an abundant harvest in southern and western counties but not quite as good in the east. Wool 9s 0½ per tod.

1413 An average harvest with a wet autumn. Wool 8s 4¼d per tod.

1414 An abundant harvest, with a dry summer. Grass was scarce, and ewes consequently gave little milk. Wool 8s 5½d per tod.

1415 Not such a good harvest. Wool, 7s 8¼d per tod.

1416 Evidently a patchy harvest, with much grain of low quality Wool 7s 4¾ per tod.

1417 An abundant harvest again. Wool 6s 7¾d per tod.

1418 A wet summer but with a fairly plentiful harvest except perhaps for wheat which was not better than average. Wool 6s 8d per tod.

1419 Apparently a fairly good harvest. Wool 6s 3d per tod.

1420 A good crop of wheat in the eastern counties but rather poor in the south and west, and barley yields, although heavy, were of low quality. Wool 8s per tod.

1421 An excellent harvest. Wool 8s per tod.

1422 Another abundant harvest. A poor fruit year. Wool 6s 7d per tod.

1423 Yet another plentiful harvest. Wool 7s 8d per tod.

1424–26 More good harvests. Wool 8s 5½d per tod.

1427 This was a wet summer, with rain intermittently from April

1st till November 1st. It ended the period of extraordinarily good harvests.

1428 Another wet year, with poor harvest. Wool 7s 1½d per tod.

1429 Another year of excessive rains during summer. Wool 8s 8d per tod.

1430 A good harvest year. Wool 7s 8d per tod.

1431 Apparently another good harvest. Wool 7s 11¼ per tod.

1432 As prices went up, this harvest was probably rather deficient. Wool 6s 9½d per tod.

1433 An excellent harvest. Wool 4s 10d per tod.

1434 Another abundant harvest. In autumn a very cold period began on November 24th, which continued until February 10th 1435, during which period the Thames was frozen over at London and as far east as Gravesend. In Scotland it is said that the frost was so severe that ale and wine were sold by weight and then melted in front of the fire. Wool 5s 4d per tod.

1435 Apparently a good harvest in spite of the early frost. Wool 4s 10d per tod.

1436 Apparently the harvest was no better than average. Wool 6s 1¼d per tod.

1437 Evidently a poor harvest. Plague was prevalent in Scotland. Wool 4s 4d per tod.

1438 A year of famine. Rain was almost continuous throughout summer and was accompanied by low temperatures and dull weather. The harvest was consequently very poor. Rye was imported from Prussia to London by the Lord Mayor and did much to alleviate suffering among the poor. Wool 4s per tod.

1439 Another stormy year with rather poor harvests. Wool 6s 5d per tod.

1440 Evidently quite a good harvest. Wool 6s 9¾d per tod.

1441 An abundant harvest. Wool 4s 8d per tod.

1442 Another excellent harvest. Wool 5s 9½d per tod.

1444 Also very good harvest. Wool 5s 4d per tod.

1445 A poor harvest year, especially for wheat. Wool 5s 6½d per tod.

1446 An average harvest year. Wool 5s per tod.

1447 A very hot summer with no better than average harvest. Wool 4s 4d per tod.

1448–9 Apparently average harvests. This sort of yield continued until 1454. Wool 5s 1d–3s 9½d per tod.

1454 An exceptionally abundant harvest. Wool 3s 4d per tod.

1455 Apparently an average harvest. Wool 3s 4d per tod.

1456 A rather deficient harvest. Wool 5s per tod.

1457 Another poor harvest. Wool 4s per tod.

1458 Evidently a very good harvest. Wool 4s 4¼d per tod.

1463 After a period of apparently average harvests this year brought abundant yields. Wool 7s per tod.

1464–77 Apparently a succession of very good years, with yields almost always above previous averages. Wool 4s 10d per tod.

1473–4 Had very hot summers. Wool 5s 2¾d per tod.

1477 Apparently a wet summer, but with much hot weather. The harvest this year was not so good. Wool 6s per tod.

1478 Another wet harvest with corn of inferior quality. Wool 4s per tod.

1479 Apparently a very good harvest but plague caused heavy mortality during the autumn.

1480 Evidently an average harvest. Wool 3s 8d per tod.

1481 A wet summer with a deficient harvest. Wool 4s 4d per tod.

1482 Another cold, wet summer with a rather poor harvest. Wool 5s 7d per tod.

1483 Another poor harvest, especially in the West. A disease known as sweating sickness caused many deaths in England. Wool 5s 2d per tod.

1484–6 Evidently fairly good harvests. Another unidentified disease in the autumn caused heavy mortality among human beings. Wool 5s 4d per tod.

1487–97 A series of apparently good harvests, except for 1490, when the yields were deficient. Wool 4s per tod.

1498 A year of drought but apparently a fairly good harvest. Wool 8s per tod.

1500 Harvest no better than average, and much plague among the animals and men. Wool 6s 8d per tod.

1501 A year of some scarcity.

1502 Another bad harvest. Wool 4s 8½d per tod.

1503 A poor harvest, especially in the West.

1504 A fairly good harvest, except in the West, where it was poor.

1505 An average harvest.

1506 The early part of the summer was wet but afterwards the weather seems to have brightened up and a reasonable harvest resulted. Wool 4s 10¼d per tod.

1507 Apparently quite a good harvest. More mortality caused by the return of the sweating sickness. Hops were introduced from Flanders about this time. Wool 4s 3d per tod.

1508 An abundant harvest. Wool 3s 1d per tod.

1509 Another plentiful harvest. The price of wheat this year is said to have been the lowest for more than two centuries. Wool 4s 8d per tod.

1510 Another excellent harvest. Plague was still prevalent in much of the country. Wool 6s 6¾d per tod.

1511 Apparently a poor harvest with a consequent rise in the price of corn. Wool 7s 7d per tod.

1512 A year of scarcity.

1513 Conflicting evidence about this year. July is said to have been excessively hot, but later in the summer there seems to have been much wind, rain and cool weather. However, the harvest appears to have been reasonably good. Much mortality occasioned by plague in the early part of the year. Wool 4s 11½d per tod.

1514 Apparently a good harvest. Wool 6s 8d per tod.

1515 Another good harvest. During winter the Thames was frozen over in London. Wool 6s 6¾d per tod.

1516 A year of drought with apparently a rather poor harvest. Evidently there were quite hard frosts in May, and disease caused much mortality among sheep. Wool 7s 7¾d per tod.

1517 Evidently an excellent harvest. Wool 7s 2½d per tod.

1518 A good harvest.

1519 A poor harvest with low yields. More mortality from the sweating sickness. Wool 7s 4d per tod.

1520 Apparently a very poor harvest and the price of wheat rose to the highest level since 1438. An exceptionally heavy storm on June 18th. Wool 6s 1d per tod.

1521 A slightly better harvest, except in the West. In this year a decree was passed against the enclosure of land as sheepwalks, this having resulted in a decay in husbandry and arable farming. Wool 4s 3½d per tod.

1522 A fairly good harvest.

1523 A great gale early in November was followed by exceptionally severe period of frost.

1524 Apparently a good harvest again this year.

1525 Another good harvest. Wool 4s 5½d per tod.

1526 Evidently a rather poorer harvest. Wool 4s 3d per tod.

1527 Recorded as a famine year. The year began with long periods of rain which caused considerable flooding and drowned many beasts in the fields. After the middle of January fine weather prevailed until April 12th from which time there was continuous rain until June 3rd. A very poor harvest resulted. Wool 7s 5½d per tod.

1528 Another bad year. Excessively wet weather in spring prevented much corn being sown at the proper time. Grain was imported from Germany in autumn. Wool 6s 3½d per tod.

1529 Yet another deficient harvest.

1530 A rather better harvest but yields were apparently not above average. Early in November a great gale did much damage in the eastern counties and caused much flooding.

1531–3 Harvest no better than average. Wool 5s 4d per tod.

1534 A period of severe frost from November 1533 till February 1534, with the river Thames frozen in London and to some miles below Gravesend. Then good harvests followed. Wool 6s per tod.

1535 A poor harvest, evidently with much rain, for floods are recorded a second time. Wool 8s 2d per tod.

1536 An abundant harvest, especially in the West.

1537 Another good harvest with corn prices correspondingly low.

1538 A very hot, dry year with quite a good harvest.

1539 Another very hot summer with good harvest.

1540 Great drought from February till the middle of September, with high temperatures on many occasions. The summer was exceptionally early, with cherries ripe by the end of May, and by June 25th farmers were in the middle of harvest. There was plentiful harvest of both corn and fruit. By the end of the summer the drought had caused many streams and wells to dry up, so that cattle died from lack of water.

1541 Another year of drought, with many small rivers drying up and many cattle dying from lack of water. The River Trent became a trickling stream, and sickness, characterised by dysentery, attacked many people. The harvest seems to have been rather deficient.

1542 Apparently a wet summer, but an abundant harvest.

1543 A fair harvest, better in the West than elsewhere, but much sickness and mortality among sheep and cattle.

1544 Evidently another rather poor harvest for the price of corn rose rapidly.

1545 A severe winter followed by famine conditions until harvest. The harvest also was deficient, and prices rose correspondingly.

1546 Much distress in the early part of the year, owing to the previous year's bad harvest, but a better harvest alleviated conditions.

1547 An abundant harvest. A very cold spell at the end of the year. Wool 9s 4d per tod.

1548 A good harvest.

1549 Apparently another deficient harvest, for the prices of corn continued high.

1550 Yet another bad harvest.

1551 Apparently a rather better harvest, as prices fell a little. The sweating sickness was again causing mortality among human beings. Wool 20s per tod.

1552 Evidently a better harvest because the price of wheat fell by half. A law was passed against profiteering in corn.

1553 A fairly good harvest.

1554 Another rather poor harvest, though conditions seem to have varied locally.

The export of wheat without a license was prohibited when the price rose to about 6s 8d per quarter, and this year the price was 18s 8¼d rising at one period to 32s.

1555 Another very wet year with consequent flooding and poor harvests.

1556 A famine year in England, following the wet and deficient harvest of 1555. 'All the corn was choked and blasted. The harvest excessive wet and rainy.' Heavy human mortality in this and the next two years, perhaps through influenza.

1557 Evidently an abundant harvest this year.

1558 A stormy summer but apparently a good harvest. Scarcity of labour accounted for much loss of corn. Wool 9s per tod.

1559 Evidently an average harvest. Wool 15s 8d per tod.

1560 Another poor harvest, to judge from the rising prices. Wool 18s per tod.

1561 A fairly good harvest.

1562 A variable harvest according to locality.

1563 Evidently a stormy summer but a rather better harvest than in recent years. The export of grain was allowed again, and in the following year wheat prices fell.

1564 The year started with very severe frosts, the Thames being frozen over at London Bridge, and football and other games were played on the ice. Then a sudden thaw caused serious flooding. Evidently the summer was wet, for floods were also reported in various places in September.

Apparently an average harvest.

1565 Evidently a rather poor harvest.

1566 A fairly good harvest, especially in the West.

Severe penalties enacted against the export of sheep.

1567 Evidently a rather better harvest for prices of corn fell.

1568 A dry summer with average harvest. There was a considerable scarcity of hay and green fodder and cattle suffered in consequence. The following winter was exceptionally hard.

1569 Apparently an average harvest.

1570 A rather better harvest, but a wet and stormy autumn. Wool 16s per tod.

1571 Apparently a rather poor harvest.

1572 A severe winter was followed by a late spring with the wind persistently in the north and east and much frost and snow. The harvest was deficient.

1573 Heavy storms in early summer. The harvest appears to have been poor, though less so in eastern counties than elsewhere. Prices were high, that of wheat rising to an average of 26s 3¾d per quarter.

1574 Much better harvest, and the price of wheat fell to about half the average for the previous year.

1575 Evidently not a good harvest as the price of wheat rose considerably in the following year. Wool 20s per tod.

1576 A rather better harvest but not abundant.

1577 A fairly good harvest. Wool 14s per tod.

1578 An average harvest apparently. An unusual snowfall

occurred in Wiltshire on May 6th, when at Broadchalke the mourners had to dig a way to the Church porch to allow a corpse to pass.

1579 A wet and stormy year. There was an exceptionally heavy snowstorm in early February. When the snow suddenly melted, a few days later, considerable flooding resulted. Another heavy snowstorm occurred in London on April 24th. Then, in autumn, gales and storms caused flooding in many parts of the country. Apparently the harvest was deficient.

1580 After a cold winter, another stormy summer, with much thunder. Apparently the harvest was deficient.

1581 Evidently another poor harvest to judge by rising prices.

1582 Another wet summer with heavy rain and thunderstorms in August. The harvest was disappointing.

1583 A fairly good harvest.

1584 Another very wet summer with great damage to hay and corn by thunderstorms in June and July.

1585 Evidently an excellent harvest, as the price of corn fell quite dramatically from 24s a bushel in May to 9s in September.

1586 Another stormy summer, with its climax on October 8th when a tremendous storm raged throughout the country and did an enormous amount of damage. The harvest was very poor and corn prices soared, to as high as 64s a quarter in the following year.

1587 Another year of rain, floods and poor harvests.

1588 Although the gale off the south coast on May 30th which disabled the Spanish Armada, brought much rain, the summer was finer than previous ones, and an abundant harvest was reaped.

1589 An average harvest, better in the West.

1590 Another average harvest, again better in the West.

1591 A year of drought.

1592 Another dry year after a severe winter.

1593 Apparently a poor harvest, as wheat prices rose considerably in the following year.

1594 A year of scarcity, with wheat rising to 56s a quarter. Rain fell almost incessantly from early May to July 25th. Harvest was poor.

1595 Apparently another poor harvest.

1596 The series of bad harvests continued, and wheat prices at Exeter rose to 62s 9¼d per quarter. Famine throughout Europe. Food riots in England.

1597 This is recorded as a famine year with a poor harvest and the average price of wheat rising to 92s a quarter.

1598 Apparently a better year for the price of wheat dropped considerably.

1599 Another good harvest year.

1600 Wheat 36s per quarter.
Another good harvest year. Heavy snow-fall on April 24th.

1601 Wheat 34s 10d per quarter.
Another good harvest year.
Hugh Platt estimates that a yield of 4 quarters of wheat per acre was about average and that some farmers were obtaining 6 or 7 quarters.
A plague of caterpillars in Pembrokeshire in June.

1602 Wheat 29s 4d per quarter.
A good harvest year.

1603 Wheat 35s 4d per quarter.
A good harvest year.

1604 Wheat 30s 8d per quarter.
Another good harvest year.
The ceiling price for wheat, above which the export of

wheat was prohibited was fixed at 26s 8d. Rye, beans and peas were not allowed to be exported when the price was above 15s a quarter; and barley was not allowed to be exported when the price was over 14s a quarter.

1605 Wheat 35s 10d per quarter.

A good harvest year.

1606 Wheat 33s 10d per quarter.

Another good harvest year.

Severe floods by the Severn in January, many people and cattle being drowned between Gloucester and Bristol. Widespread floods also in many places in the West of England and in Wales. The river Brue burst its banks at Burnham in Somerset, flooding 30 villages and destroying many cattle. At Kingston Seymour, in Somerset, the flood water was as much as 12 feet deep. At least 2,000 persons reported drowned in Monmouthshire, Glamorgan and Somerset in this disaster. At the same time heavy flooding occurred in the Fens and in Romney Marsh.

1607 Wheat 36s 8d per quarter.

A very hot summer, with drought. A poor harvest, especially in the West.

A long period of severe frost began on December 5th and continued until February 14th, 1608. Many rivers were frozen over, ice being thick enough to bear a horse and cart. Ice stopped the working of many mills. In London fires were lighted on the ice on the Thames in December.

1608 Wheat 56s 8d per quarter.

Another bad harvest.

The price of wheat rose very rapidly during the winter, probably because of the severe weather.

Exceptionally severe frost in January. The Thames was frozen over in London. The thaw came about February 14th.

A great gale caused much havoc at Beverly in Yorkshire.

1609 Wheat 50s per quarter.

A much better harvest in the West.

1610 Wheat 35s 10d per quarter.

An average harvest.

A exceptionally heavy thunderstorm in Lincolnshire on July 3rd, with much damage to crops.

1611 Wheat 38s 3d per quarter.

An average harvest.

1612 Wheat 42s 4d per quarter.

Another average harvest.

1613 Wheat 48s 8d per quarter.

A poor harvest.

A severe thunderstorm in southern England on the evening of June 26th, it being exceptionally violent at Southampton.

1614 Wheat 41s $8\frac{1}{2}$d per quarter.

Harvest about average.

1615 Wheat 38s 8d per quarter.

An unusually heavy snowfall began on January 16th. This snow was exceptionally deep, covering the road and fields to the tops of the gates and hedges. It was followed by further snow until March 12th, and in Derbyshire there were also at least ten snow-storms during April. A further snow-storm occurred in March. This winter snow was accompanied by severe frost.

The subsequent thaw caused severe flooding particularly in Yorkshire and Lincolnshire, The river Ouse overflowed its banks at York and flooded the streets, destroying many bridges. On the flat country around Boston many sheep were drowned.

The summer was warm and dry, drought developing during July and August. Hay and corn became very scarce, the price of hay at Leeds rising to 80s a load.

1616 Wheat 40s 4d per quarter.

A drought developed during the summer.

Large-scale enclosures of common land began.

1617 Wheat 48s 8d per quarter.

A rather poor harvest.

1618 Wheat 46s 8d per quarter.

A much better harvest, except in the West.

1619 Wheat 35s 4d per quarter.

The growing of tobacco in England prohibited by law.

1620 Wheat 30s 4d per quarter.

A severe period of frost in winter froze the Thames sufficiently for a Frost Fair to be held on the ice. Snow fell for 13 days continuously in Scotland. It is reported that on Eskdale Moor only 45 sheep were left alive out of a total of 20,000.

The river Severn overflowed in Worcestershire on November 29th, causing much damage and many casualites.

1621 Wheat 30s 4d per quarter. This price was so low that there was considerable anxiety and even distress among farmers.

The price of land fell by about 15%.

1622 Wheat 58s 8d per quarter.

A poor harvest.

The wool market reached the nadir of its depression, the lowest wool prices for many years being recorded. Wool in Lincolnshire fetched 10s per stone as against 14s in 1618. A very hard winter.

1623 Wheat 52s per quarter.

The ceiling price for wheat above which the export of wheat was prohibited was fixed at 32s.

1624 Wheat 48s per quarter.

1625 Wheat 52s per quarter.

A very severe winter was followed by violent outbreaks of plagues during the summer.

1626 Wheat 49s 4d per quarter.

John Lawrence estimates that one-third of the cultivated land in England and Wales was in common fields.

1627 Wheat 36s per quarter.

Prices given for poultry are as follows:

Turkey cocks	4s 6d	Goose	2s
Turkey hens	3s	Capon	2s 6d
Cock pheasant	6s	Pullet	1s 6d
Hen pheasant	5s	Rabbit	8d
Partridge	1s	A dozen pigeons	6s

A severe gale from the south on January 28th. On April 9th a violent thunderstorm occurred in southern England.

1628 Wheat 28s per quarter.

This was a very wet year, with much disease in sheep.

Flooding occurred during autumn.

1629 Wheat 42s per quarter.

Severe floods in February.

1630 Wheat 55s 8d per quarter.

A very dry summer.

Formation of a Company of 'Adventurers' to undertake the drainage of a large area of Fenland, led by Francis, Earl of Bedford. Reclamation of the Fens began, employing the Dutch engineer, Vermuyden.

1631 Wheat 68s per quarter.

This high price followed an extraordinary drought with a general scarcity of corn.

1632 Wheat 53s 4d per quarter.

A very wet summer.

1633 Wheat 58s per quarter.

A great flood occurred, with numerous casualties, at Newcastle-on-Tyne.

1634 Wheat 56s per quarter.

The Thames was frozen over in the winter frost.

1635 Wheat 56s per quarter.

Heavy snow, accompanied by frost occurred in southern England in January, followed by a thaw which produced a great flood. On February 5th severe flooding occurred in Salisbury, the water standing a foot deep in Salisbury Cathedral.

The practice of constructing water meadows began in Wiltshire about this time.

1636 Wheat 56s 8d per quarter.

A very early spring.

An exceptionally long drought began about March 1st and lasted for most of the summer.

Plague was widespread.

1637 Wheat 53s per quarter.

1638 Wheat 57s 4d per quarter.

A damp and wintry autumn, with gales and tornadoes in the West Country on October 31st.

There was an earthquake at Chichester, which did considerable damage, at the end of the year.

Gabriel Platts patented an implement for setting corn in drills.

1639 Wheat 44s 10d per quarter.

The practice of burn-baking land was introduced to Wiltshire from Flanders by a Mr Bishop of Merton.

1640 Wheat 44s 8d per quarter.

A very violent thunderstorm occurred in Cornwall on Whit-Sunday.

1641 Wheat 48s per quarter.

Severe floods in the Fens in April.

1642 Wheat 60s 2d per quarter.

A heavy gale in the Midlands on August 27th.

1643 Wheat 59s 10d per quarter.

A severe storm occurred in southern England on November 6th, followed by unseasonably cold weather with snow.

1644 Wheat 61s 3d per quarter.

Severe snowfall from January 39th to February 7th.

An exceptionally heavy thunderstorm occurred in the Midlands on May 15th and 16th, during which enormous hailstones, bigger than walnuts, fell.

1645 Wheat 51s 3d per quarter.

A very cold and stormy February, followed by great floods in March.

The summer was warm.

1646 Wheat 48s per quarter.

A very cold January with much snowfall, followed by a hot summer.

1647 Wheat 73s 8d per quarter.

An Act of Parliament was passed to prohibit the export of wool.

1648 Wheat 85s per quarter.

'A most exceeding wet year, neither frost nor snow all the Winter for more than 6 days in all . . . prodigiously wet summer, very cold.' *John Evelyn.*

In consequence of this wet weather there was a great scarcity of grain, and the prices went up accordingly.

Cattle also suffered a great deal from disease.

1649 Wheat 80s per quarter.

The Thames was frozen over in London in January.

A very hot spell in June.

Corn scarce, this year being recorded as a famine year.

1650 Wheat 76s 8d per quarter.

Sir Richard Weston published his book, *Discours of Husbandrie.*

1651 Wheat 73s 4d per quarter.

1652 Wheat 49s 6d per quarter.

A very hard and prolonged period of frost in January, followed by a spring drought.

This broke in southern England with a violent thunderstorm on May 25th, accompanied by severe hail.

The summer continued hot and dry, it being recorded as the driest ever known in Scotland.

An eclipse of the sun occurred on April 29th.

W. Blith published his book, *English Improver Improved.*

1653 Wheat 35s 6d per quarter.

1654 Wheat 26s 8d per quarter.

An exceptionally hot spell in early July.

1655 Wheat 33s 4d per quarter.

The winter was exceptionally cold in Scotland.

1656 Wheat 43s per quarter.

Prolonged frost occurred in mid-January, and a snow-storm covered the hills in southern England with snow on May 1st.

The following summer was unusually hot and dry.

1657 Wheat 46s 8d per quarter.

Exceptionally heavy rain in southern England on August 21st.

A fairly severe earthquake occurred at Bickley in Cheshire on July 8th.

Snow which fell in England on December 11th is reported to have lasted until March 21st 1658.

1658 Wheat 65s per quarter.

The winter, in the early part of the year, is reported to have been the severest within living memory in England.

An unusual storm of hail and rain occurred on June 2nd.

The summer was cold with continuous northerly winds.

On August 18th a gale blew up which continued for two days and did much damage, especially to fruit trees in the south-west of England.

On September 3rd another very violent gale occurred. This was the day of Oliver Cromwell's death, and the coincidence attracted considerable attention and comment.

1659 Wheat 66s per quarter.

A very severe gale occurred on December 8th, particularly in northern England.

1660 Wheat 56s 6d per quarter.

The ceiling price for wheat above which the export was prohibited was fixed at 40s a quarter.

The corresponding price for rye, and beans was 24s. For barley and malt 20s, and for oats 16s per quarter.

A very pleasant spring.

1661 Wheat 70s per quarter.

An unusually mild winter with not much rain.

Samuel Pepys noted on January 21st that the ways were

dusty and the rose-bushes full of leaves.

On February 18th a great storm, with violent winds and thunder, lightning and hail, occurred in London.

A dry summer.

The river Derwent dried up in many places.

The autumn was wet and misty.

1662 Wheat 74s per quarter.

January was unusually warm with much rain.

Samuel Pepys records that on January 15th a general fast was declared to intercede for colder weather.

On February 17th a tremendous thunderstorm occurred, occasioning much damage and some casualties.

This is recorded as a famine year.

1663 Wheat 57s per quarter.

The ceiling price for wheat, above which its export was prohibited, was fixed at 48s.

The corresponding price for barley was 28s, for oats 15s 4d, for rye 32s, for beans and peas 32s.

This was an exceptionally cold and wet summer.

On July 21st Parliament kept a fast because of the unseasonable weather.

On August 28th a severe frost occurred.

Samuel Pepys described an extraordinarily fierce thunderstorm at Northampton on May 8th.

Severe mortality caused among sheep by disease known as rot.

1664 Wheat 40s 6d per quarter.

A severe and prolonged frost lasted from the beginning of the year to early March.

A further exceptionally cold spell began in the third week of December.

John Forster urged farmers to grow potatoes as a field crop.

1665 Wheat 49s 4d per quarter.

A cold and frosty January and February.

Another cold spell began on October 19th, with heavy rain in London.

November was a month of unusual wind and rain with

the exceptionally low barometric reading of 931mm, recorded in London.

A great gale occurred on October 25th.

A period of severe frost occurred in southern England from December 2nd to 7th, and a further period began on December 21st.

The Thames was frozen over in London on December 13th.

The great plague of London.

1666 Wheat 56s per quarter.

The year began with a severe frost, which ended on January 6th. After that date boats were able to move again on the river Thames and a mild January followed.

In a severe gale on February 3rd houses were blown down in London.

A very hot and dry summer followed, the main heat-wave and drought beginning on June 27th.

Severe thunderstorms occurred at Andover on May 12th and at Oxford on July 17th. Also in Suffolk on the latter date with very large hail-stones.

There was a short period of heavy rain beginning on December 6th and heavy hail-storms occurred in London and on the East Coast on July 26th and 27th. Thereafter August and September were unusually dry and warm.

The great fire of London broke out on September 12th being fanned with strong east winds, and the excessive heat and drought of the previous months were one of the causes of the tremendous damage done.

The drought broke on September 19th, after which the autumn was exceptionally wet and rainy.

An Act of Parliament was passed ordering burial to be made in woollen fabric, as a measure to assist the wool trade.

1667 Wheat 36s per quarter.

There were hard frosts at the beginning of January, and March was a mixture of severe frost, snow and very strong winds.

The spring was late, with hardly a leaf to be seen in early April.

On June 11th a very dry spell began which lasted until mid-August, with exceptional drought in June and July.

1668 Wheat 40s per quarter.

An exceptionally heavy rain-storm in Warwickshire on July 24th.

1669 Wheat 44s 4d per quarter.

This was an unusually dry year with a scarcity of water developing in many parts of the country.

At the end of the year a very severe spell of frost occurred, with heavy snowfall on and after December 26th. The barometer reached the unusually high level of 30.6 ins at Bristol on December 14th.

Publication of Worlidge's, *Systema Agriculturae,* which includes an illustration of a horsedrawn drill which cut a furrow and sowed seed in one operation.

1670 Wheat 41s 8d per quarter.

A very violent gale occurred at Braybrook, Northamptonshire, on October 13th.

An exceptionally dense fog developed in London on December 15th, and the year ended with intense frost.

1671 Wheat 42s per quarter.

A very wet and stormy January.

A severe storm on September 13th.

An ice-storm occurred in the West Country on December 9th, 10th and 11th. The rain froze as it fell and coated everything with a film of ice. An ash-tree branch which was measured had ice on it 5 inches in circumference, and carried a weight of 16 lb of ice although its own weight was only $\frac{3}{4}$ lb. Many roads were blocked by fallen trees collapsing under the weight of ice. This storm reached as far as Oxford. It was followed by an unusually warm spell, with bushes and flowers blooming.

1672 Wheat 41s a quarter.

The year started with a wet winter, with an exceptionally heavy snowstorm on February 3rd.

1673 Wheat 46s 8d per quarter.

This was a famine year, the harvest being very poor.

Poor people in the West Country were obliged to make their bread of peas and beans, through scarcity of wheat.

Early in the year great numbers of cattle and sheep died in continuous wet weather, the mortality being so great that Orders of Sessions were made compelling owners of deceased cattle to bury the bodies.

Heavy floods in December.

1674 Wheat 68s 8d per quarter.

The year began with severe floods.

On March 8th began the unprecedented blizzard along the Scottish borders which lasted for 13 days. This blizzard was long remembered and caused tremendous loss of sheep; also of horses and cattle.

This is recorded as a famine year with a very poor harvest.

The wet weather caused great floods in the Midlands on May 7th and 8th.

Exceptionally severe gale occurred in Scotland on December 21st, uprooting whole forests.

1675 Wheat 64s 8d per quarter.

The summer was exceptionally dry and the harvest consequently poor.

An earthquake occurred at Alrewas in Staffordshire on January 4th.

A period of unusual heat lasted from June 19th to July 1st.

1676 Wheat 38s per quarter.

A dense fog recorded on October 31st.

Heavy snow from December 10th onwards.

1677 Wheat 42s per quarter.

In this year regular rainfall observations began for the first time in England. They were made at Townley in Lancashire, where the rainfall for the year was 43.6 inches.

Wool prices were still very low, being 6s 8d per stone in Lincolnshire.

1678 Wheat 59s per quarter.

A very hot dry summer and autumn. Rainfall at Townley 42.7 inches.

1679 Wheat 60s per quarter.

A very hot dry year, with next to no rain from May onwards.

The following autumn was wet.

Rainfall at Townley 38.2 inches.

1680 Wheat 45s per quarter.

November was unusually dark, cloudy and snowy, and severe weather followed.

An earthquake occurred in Somerset on January 4th.

The beginning of a further depression in the wool industry, when wool prices again declined.

Rainfall at Townley was 44.3 inches.

1681 Wheat 46s 8d per quarter.

A very cold and late spring, with frost and snow still lying on the ground at the end of March.

The whole of the early part of the year until midsummer was exceptionally dry.

Rainfall at Townley 33.3 inches.

The amount of rain which fell in 1681 was only about 76% of the normal.

The Rector's Book of Clayworth in Nottinghamshire records on June 18th 'Barley found dry in ye fields, having lain so ever since sowingtime'.

It further records, 'it was a very dry and droughty year fro ye begine of April to ye 20th June, not having raynd, except on ye 7th May. But after Xtmas fell abundance of rain . . .'

A four months' spell of dry weather and easterly winds in Scotland ended on July 4th.

1682 Wheat 44s per quarter.

A warm winter and a rather wet spring and summer. April was unusually wet with much thunder.

A sudden flood occurred in March at Bradford and did much damage.

Rainfall at Townley 50.7 inches.

1683 Wheat 40s per quarter.

A severe spell of weather, which began at the end of 1682, continued till February 5th.

The Thames was frozen with ice 11 inches thick. Many birds perished.

With the thaw came tremendous damage by floods,

especially at Nottingham where the Trent bridge was almost destroyed by lumps of ice floating down the stream.

March was hot and dry, and April unusually wet. The summer was wet and cold until the middle of September, with one fine week from September 20th to 27th. Then frost began with a very high reading of the barometer.

On December 15th began a most remarkable spell of cold weather extending right across Europe to England. It is recorded that many trees, particularly oaks, split with the frost, exploding with a noise like a gun. Even yew and holly trees in some places were killed and many shrubs.

Coaches were on the ice on the Thames by the end of the year.

The severest part of this spell, however, occurred in the following year 1684.

Rainfall at Townley 37.2 inches.

1684 Wheat 44s per quarter.

The exceptionally severe spell of cold weather which began in December 1683 continued for the first eight weeks of 1684. John Evelyn gives details of the frost in London. The river was quite frozen over by the 9th January and streets of booths were set up on the Thames. By the 16th the shopping streets stretched from bank to bank, and horses, carts and coaches crossed over on the ice. Printing presses were set up to print, as souvenirs, cards giving the name of the person who bought them, the date and the year, and the fact that they were printed on the ice on the Thames.

There was some heavy mortality among fish and birds, and also among plants and trees, the intense frost splitting many large trees. In many parks deer perished.

The sea froze in several places along the coast.

London, in the tranquil and excessively cold air, was affected by dense fog.

There were temporary thaws on February 5th and 10th, but the weather did not become really mild again until towards the end of March.

This severe weather extended over the greater part of Europe, and was followed by a very dry spring.

The ensuing drought was one of the most severe within living memory and did not break until about August 21st. Heavy rain followed, but the weather still continued hot.

The autumn likewise remained warm until November 2nd, when there was a sudden change to a wintry climate with a tempest of snow and cold rain, followed by frost. This cold weather lasted until Christmas, being distinguished by an exceptionally severe blizzard in southern England on December 23rd. Many people perished in the snow.

Rainfall at Townley was 34.1 inches, which was only 79% of the normal, making this one of the driest years on record.

1685 Wheat 46s 8d per quarter.

Another very cold winter, with the Thames frozen over in London.

The spring was excessively dry and John Evelyn records that a plague of caterpillars destroyed many fruit trees.

The drought continued until about the middle of June, after which rain fell liberally.

Summer, autumn and early winter were warm and wet.

Rainfall at Townley 37.8 inches, of which the greater part fell in the latter months of the year.

1686 Wheat 34s per quarter.

The year started with an exceptionally wet and mild winter, and so it continued.

Rain, storms and generally foul weather with high temperatures, were recorded for early June. This weather continued during July, and the autumn was unusually rainy.

Rainfall at Townley 50.4 inches, August having an exceptional rainfall of 8.7 inches.

1687 Wheat 25s 2d per quarter.

This proved to be a very hot, dry summer. Crops were not good, and the low prices caused considerable anxiety in the countryside.

On May 12th a hurricane occurred in London, and an earthquake was recorded in several places in England on the same day.

June was exceptionally windy.

1688 Wheat 46s per quarter.

A very late cold dry spring distinguished by prolonged easterly winds, though without much sun.

Frost was very severe in January, and the Thames was again frozen.

A display of aurora borealis was seen from Bristol on October 30th.

1689 Wheat 30s per quarter.

The year started with a prolonged frost and deep snow.

On January 11th a severe gale occurred.

The spring was warm and seasonable, but the early summer was characterized by drought, which severely checked the growth of the crops.

Summer was warm and pleasant on the whole, though a violent thunderstorm occured on July 11th.

Autumn proved unusually wet. During one wet spell in particular rain began on October 4th and continued without a break until October 10th. This caused severe flooding, especially in Norfolk.

Rainfall at Townley 48.6 inches.

1690 Wheat 34s 8d per quarter.

The winter was rainy and rather warm without much wind.

A severe blizzard occurred on the night of January 11th.

The summer was fine and warm until August 12th, when a heavy thunderstorm heralded the approach of wet, cold weather.

In autumn several violent gales and storms occured, though the weather remained fairly mild.

December was cold.

Rainfall at Townley 42.9 inches.

Wool prices were still low, with 8s per stone, the top price for best Lincolnshire wool.

The sheep and lamb population of Great Britain was estimated at about 13,000,000.

1691 Wheat 34s per quarter.

This was an unusual year with moderate amounts of rain during the spring and early summer.

During the later summer drought developed and continued throughout the autumn and early winter.

John Evelyn speaks of November as 'an extraordinary dry and warm season without frost, and like a new spring, such as had not been known for many years'.

A heavy thunderstorm occurred on August 20th and another storm, which occasioned great loss of life at sea, on September 13th.

Rainfall at Townley 31.4 inches. This was only 72% of the normal.

1692 Wheat 46s 8d per quarter.

This began a series of very adverse seasons for farmers.

January and February were frosty and dry, with a heavy snowfall on February 7th. This cold dry weather continued throughout April and early May, and the growth of crops was consequently very backward. A wind blew consistently from the east.

On February 21st a severe blizzard from the north-east swept Scotland.

On June 9th a great storm, with very heavy rain and violent wind caused considerable damage.

A period of very heavy rain and floods followed from June 19th onward and continued throughout July and August. The rainy weather came to an end on August 26th and was followed by a period of fine weather. The autumn, however, proved very cold, and much fruit failed to ripen.

A long spell of wind and gales occurred from September 25th onwards but without much rain. Oddly enough, in Ireland the year was distinguished by a summer drought.

On September 8th an earthquake was felt in London.

Rainfall at Townley 43.7 inches.

1693 Wheat 67s 8d per quarter.

The year began with a very mild winter. Although heavy snow fell on February 26th, it thawed again very quickly.

Spring was unusually wet, and much hay was spoiled during the exceptionally wet month of June. Thereafter more rain fell than usual, though there was a good harvest period in August.

Autumn also continued very wet, Corn was abundant but hardly any fruit was harvested.

The wet summer extended to France, where considerable distress was caused, in spite of corn being imported from Scandinavia.

It was also a bad year for sheep disease in England, notably liver fluke.

A heavy thunderstorm occurred in southern England on December 10th.

Rainfall at Townley 42.3 inches

1694 Wheat 64s per quarter.

Another very wet year with a poor harvest.

April was unusually dry, and the drought continued until May 13th, when it was broken by showers. After some showers in June, July proved warm and sunny, with a prospect of a magnificent harvest. However, in August wet weather set in, resulting in a great deal of waste.

A severe period of frost began on December 27th.

1695 Wheat 53s per quarter.

The year began with a long spell of severe weather. By January 13th the Thames was frozen over, and the winter, which was characterised by severe cold and snow, did not end until April 14th.

The first rain for several months fell on April 21st.

During this late spring the Arctic Sea ice came much farther south than usual.

The summer proved wet, and by August it was cold as well. Night frost was recorded in Surrey on August 21st.

The harvest was difficult with much rain, heavy storms and relatively cold weather.

A period of rain and gales lasted until October 12th but the weather got progressively milder. The harvest is recorded as 'deficient'.

There was considerable migration of population from Scotland into Ireland because of the excessive price of corn in Scotland.

1696 Wheat 71s per quarter.

This also was a bad harvest year with a very wet summer. It began with wet though rather mild winter. March was cold and frosty with continuous north and east winds.

June was exceptionally rainy and cold, and heavy rain also fell during July. August, however, provided moderately good harvest weather, the period of exceptional rain and gales ending about August 15th.

September provided good harvest weather, at least until September 18th, and a fair harvest was gathered, though it was better in England than in Scotland, where there was a considerable scarcity of good grain.

A very stormy and wet autumn lasted from September 18th to November 13th, when a short spell of very severe frost set in.

From November 21st to December 10th much rain fell with strong westerly winds, but about December 11th a period of frost and snow began, with heavy frost and snow from December 27th to December 29th.

Early in the year the temperature in London fell to 9° below zero on January 26th.

This was a year of rising prices to match the poor harvest.

Gregory King estimated the total acreage in England and Wales at 39 million of which 11 million were arable, 10 million meadow and 10 million pasture. The average yearly rent of the arable land was 5s 10d and of the pasture land 5s. Of the arable land 10 million acres were devoted to corn, beans and vetches, and one million to flax, hemp, saffron and miscellaneous crops.

It was further estimated that there were in the country 640,000 horses and donkeys, 4.5 million cattle, 11 million sheep and 2 million pigs. The human population of England and Wales was estimated to be 5.5 million, of whom 1,400,000 lived in towns and 4,100,000 in the countryside.

1697 A very cold mid-winter spell about February 21st. The frost during this period was severe and caused numerous casualties.

A severe north-east gale occurred on January 8th.

This severe winter was followed by a very wet spring, and the summer produced another deficient harvest.

Exceptionally heavy thunderstorms occurred on April 29th, May 4th and June 6th, with hailstones of enormous size, killing fowls, rooks and hares.

Autumn was cold, and a week of tremendous storms caused casualties around October 3rd.

Rainfall at Townley 38 inches.

1698 Wheat 68s 4d per quarter.

The year began with a stormy winter with much ice and snow lasting well into spring. A great snowstorm occurred on February 14th, with snowdrifts several yards deep.

During February the wind blew consistently from the north-east, and the weather was very cloudy. Ice was 8 inches thick on the sea coast off Suffolk on February 8th.

A warm spell intervened at the end of March, but by April 22nd it was snowing again.

On May 1st a very heavy snowfall with severe frost occurred in Yorkshire.

On May 8th thick snow lay all over England. On May 13th snow fell in several places including London, with much damage to crops, particularly fruit.

On May 9th heavy snow fell in Shropshire.

On May 15th the woods were deep in snow, like winter, and severe frost continued.

On June 3rd a thunderstorm occurred with enormous hailstones covering the ground 3 inches deep.

This spring was recorded as the latest and most backward for 47 years.

The weather broke with a thunderstorm on July 9th, and this was followed by excessive rain for a week or two.

Frost occurred in the middle of August, and the second half of the month gave a short spell of dry weather. It was, however, interposed between two long wet spells which spoilt most of the harvest.

Exceptionally heavy rain fell on August 6th, the rain drops being recorded as enormous.

On some farms the first wheat was not cut until the middle of September, and much barley lay on the ground in December. Some of the harvest in the north was still ungathered at Christmas, and in Scotland corn was reaped in January 1699, the snow being beaten off. Bread made from it fell to pieces and tasted sweet, like malt.

October was wet with exceptionally cold nights, and the frequent rain throughout the summer caused flooding during late autumn, particularly at Derby.

The harvest was, quite naturally, recorded as 'deficient'.

Rainfall at Townley 41 inches, but very little of this fell in January, February, May and June.

1699 Wheat 64s per quarter.

A better year for agriculture.

After seedtime a long drought occurred, but the summer was dry and the harvest uneventful.

The Farmers' Almanac recorded another deficient harvest.

It is recorded also as a good year for aftermath grass.

Severe storms occurred on February 7th and February 9th and also on March 26th.

A long spring drought broke on June 7th, and on June 25th John Evelyn records exceptional heat, which lasted apparently until July 23rd.

A very mild and pleasant autumn was broken by a exceptionally sharp frost on October 21st.

On November 29th London was affected by a very dense fog.

December also was sunny and mild.

During June one extraordinarily cold wet day and night just after shearing caused the death of many sheep.

1700 Wheat 40s per quarter.

This apparently was a year with which the chroniclers can find little fault.

It started mild and calm, though with a setback by severe frost on January 25th.

Warm, springlike weather lasted throughout most of February and all of March and April, although May was cold and wet, and the rest of the summer proved sunny and entirely favourable to agriculture.

In this year there were estimated to be 330,000 farmers and 573,000 agricultural day labourers in England and Wales.

1701 Wheat 37s 8d per quarter.

This was another excellent year, with a magnificent harvest.

It began with a very severe snowfall on January 4th, which melted almost immediately.

A severe storm followed by frost did a great deal of damage on January 19th, but thereafter the spring was relatively uneventful, though distinguished by drought rather than rain. This drought broke with south-westerly winds and rain on May 27th.

August was particularly hot.

1702 Wheat 29s 6d per quarter.

Yet another hot dry summer.

March was a dry month, and very little rain fell from April 23rd to May 29th. Thereafter heavy showers occurred, but in general there was a great scarcity of hay and grass.

The summer in general was warm, and the harvest was gathered without difficulty.

An unusual heavy storm occurred on February 3rd and 4th.

1703 Wheat 36s per quarter.

This was a wet year with a late harvest and consequently low yields, though there was an abundance of grass.

It started with the aftermath of the 1702 harvest, corn being so cheap in the early part of the year that farmers were unable to meet their commitments.

April, May, June and July proved excessively wet. Both May and June had far more than their normal quota of rain. July had sunny periods but interspersed with heavy showers and on the whole was cold and wet.

The autumn was distinguished by extraordinary storms and wind, the most oustanding being a hurricane on November 26th and 27th. This was the climax to a period of very windy weather which had lasted for 18 days before the hurricane.

Daniel Defoe records that the worst of the storm began on Wednesday the 24th and lasted till the following Wednesday. People were afraid to go out-of-doors and were afraid to go to bed. Eddystone Lighthouse was washed away. Twelve warships were sunk off the coast. The Bishop of Bath and Wells was killed in his bed when a chimney stack collapsed on him. Many houses, barns and trees were blown down, and casualties were heavy. In this gale, damage, in London alone, is estimated to have reached £2,000,000. 8,000 people lost their lives in floods of the Severn and Thames, and the coast of Holland. In Kent 110 houses and barns were destroyed, and in one place 16,000 sheep were drowned.

This at the time was reckoned to be the most violent gale ever recorded in England, but it was apparently surpassed in southern England by another on December 7th and December 8th. Other violent gales occurred on December 27th and 28th.

1704 Wheat 46s 6d per quarter.

On the whole this was a relatively dry and uneventful year.

Spring was pleasant and favourable, resulting in plenty of grass.

A drought occurred in June and July, and the grass was burnt up. Autumn also was very dry.

1705 Wheat 30s per quarter.

Another dry season.

January started very wet, but hardly any rain fell from from April to 2nd July.

June in particular was very dry and hot.

A good harvest was gathered under almost ideal conditions.

1706 Wheat 26s per quarter.

Yet another dry summer.

Following a dry spring very little rain fell at all over the whole country for the three months before harvest. However, a violent storm with excessively heavy rain occurred in north Wales on July 16th and 17th, causing flooding and much loss of hay and corn.

Apart from this the year was uneventful.

1707 Wheat 28s 6d per quarter.

Another hot dry summer, though with showery weather in harvest.

An early dry period occurred from March 12th to May 22nd, with hot days and cold nights.

Although it was an excellent seedtime, much of the spring corn did not germinate till June.

July 19th was an exceptionally hot day in England causing many deaths.

Wheat was much damaged by blight.

In early winter a wet spell set in.

J. Mortimore published his book, *Whole Art of Husbandry*.

1708 Wheat 41s 6d per quarter.

On the whole a cold wet spring and summer.

A great snowstorm occurred on January 24th.

After a late and rainy spring, June was distinguished by frequent heavy rains.

A hoar frost occurred on June 12th.

In harvest very rainy weather prevailed from August 30th to September 20th.

A heavy thunderstorm, causing considerable floods, occurred in the West Country on September 1st.

The winter also set in cold and wet, and the Thames was frozen again.

1709 Wheat 78s 6d per quarter, reflecting the previous year's bad harvest.

This was another bad year with a very backward spring and general scarcity.

A severe period of frost began on January 7th, lasting over 50 days, during which the Thames froze.

Intense frost occurred from January 9th onwards, with heavy snowstorms and much mortality among cattle, sheep and wild birds.

April and May were also cold and wet, and crops suffered, there being a scarcity of grass.

1710 Wheat 78s per quarter.

It is recorded that by Lady-day the price of wheat had risen by 200% in two years. This is chonicled as another famine year, and the exportation of corn was prohibited for one year.

Winnowing fans were introduced in Scotland by James Meikle, after a visit to Holland,

The average weight of beef animals at Smithfield Market was 370 lb; of calves 50 lb; of sheep 28 lb; and of lambs 18 lb.

1711 Wheat 54s per quarter.

The weather of this year was the reverse of 1710, with an exceptionally dry summer but not excessively hot.

The rainfall was particularly low in the eastern counties.

The spring was wet and cold, but thereafter dry weather promoted early ripening. Wheat was cut on July 27th and barley on August 26th.

An exceptionally heavy thunderstorm, causing considerable damage, occurred at Okehampton in Devon on October 7th.

1712 Wheat 46s 4d per quarter.

A very dry spring with hardly any rain in February, March and April until the middle of May. Then an unusually hot spell set in until June 20th. There were strong winds before harvest, and the autumn and early winter proved exceptionally wet.

1713 Wheat 51s per quarter.

A dry year, with a cold spring drought followed by a dry summer.

Wheat was generally good, but barley and oats gave a poor crop.

1714 Wheat 50s 4d per quarter.

This was an exceptionally dry year, one of the driest on record. Rainfall at Upminster in Essex was only 11.2 inches for the whole year, which is less than half the average rainfall in even that dry part of the country.

Spring was cold and dry, but in summer and autumn the drought was combined with heat.

Harvest began and ended early, the yields of wheat being quite good, but those of oats and barley poor.

A serious epidemic of cattle plague (rinderpest) occurred in July.

1715 Wheat 43s per quarter.

In contrast to 1714 the summer was cold and wet. Much grain was spoiled at harvest-time by the rain, and the ensuing wet autumn caught some crops still in the field.

A very cold spell occurred in the early part of the year, the Thames being frozen over for three months. A Fair was held on the frozen river in London, and two oxen were roasted whole on the ice. Much snow fell.

A wet autumn was also followed by a severe cold spell which began on November 24th and continued till February 9th of the following year.

Total eclipse of the sun on April 22nd.

1716 Wheat 48s per quarter.

A year of drought. A cold dry spring was followed by an equally dry summer, though without great heat.

The early hay crop was poor, and the aftermath a failure. Turnips grown in the south of England were also a failure.

On September 14th the bed of the Thames lay dry above and below London Bridge. This was caused by a combination of drought, strong westerly winds and high tides.

Exceptionally brilliant displays of aurora borealis on February 23rd, March 18th, 19th and 20th.

Turnips were first grown on a field scale in Aberdeenshire by the Earl of Rothes.

1717 Wheat 45s 8d per quarter.

A spring drought was absolute from the middle of March to May 22nd. After that the summer was showery, with heavy thunderstorms at the end of July.

A fairly good year for the crops, clover growing to a great height.

1718 Wheat 38s 10d per quarter.

The summer was hot and dry, and the harvest was quite good.

The early part of the year was characterised by hard frost with considerable snowfall.

1719 Wheat 35s per quarter.

Another hot, dry summer, which followed a cold spring drought.

The hay harvest was poor.

The Michaelmas supplies of fodder were so short that cows which normally sold at £4 apiece changed hands at 30s. This drought was followed by a wet and mild autumn.

1720 Wheat 37s per quarter.

In contrast to 1719 this summer was excessively wet. During March, April and May so much rain fell that hardly a spring bite was available for cattle. The water meadows overflowed to such an extent as to be inaccessible, and farmers despaired of making any hay. The weather improved after July 18th, and the summer and autumn were characterized by an abundance of grass.

1721 Wheat 37s 6d per quarter.

A moderately good year with few distinguishing features.

A hot dry spell in the middle of June greatly assisted hay-making.

1722 Wheat 36s per quarter.

Another undistinguished year. January was frosty, and spring and early summer rather wet.

Autumn was dry, and mild weather prevailed till nearly the end of the year.

1723 Wheat 34s 8d per quarter.

A very dry year, with particularly low rainfall in February and March.

The summer was hot.

1724 Wheat 37s per quarter.

Another undistinguished year.

1725 Wheat 48s 6d per quarter.

A year of extremes. From January 13th to the middle of April an exceptionally dry spell occurred, stated to have been one of the driest ever known in England. At Wells in Somerset only 15 days with rain were recorded during those three months. This spell broke up in rain on April 12th in the west of Scotland, and rather later in the south of England.

April 25th saw the beginning of an unusually long spell of rain and gales.

Floods occurred in many parts of the country. The rain was accompanied by cold weather, resulting in poor and late ripening of fruit and vegetables.

One of the worst periods lasted from July 13th to September 4th. After this date a period of dry weather set in, with moderately high temperatures.

1726 Wheat 46s per quarter.

Considerable floods in the south of England during the early part of the year. On March 8th the Thames was 4 inches higher in London than had been known for 40 years.

An exceptional flood in Salisbury inundated the Cathedral to a depth of a foot.

An earthquake was felt at Dorchester on October 25th.

1727 Wheat 42s per quarter.

An undistinguished year.

1728 Wheat 54s 6d quarter.

Heavy rain and great floods occurred during the early part of the year, but the summer was dry, especially in the West.

1729 Wheat 46s 10d per quarter.

A year marked by an unusual number of thunderstorms and great winds.

January and February was a period of hard frosts, lasting for about nine weeks.

On May 20th tornadoes occurred in Sussex and Kent. In June heavy thunderstorms were recorded. In July also there were thunderstorms in many parts of Europe as well as in England.

In November a whirlwind did much damage in Wiltshire, uprooting trees and taking the roofs off houses in the village of Barford St Martin.

Harvest reasonably good.

1730 Wheat 36s 6d per quarter.

This year began a series of four unusually dry years, lasting till June, 1734. August was particularly dry. Harvest excellent.

In this year Viscount Charles Townshend, having lost his position as Prime Minister, retired to his estate at Raynham in Norfolk and began his notable land improvement schemes.

1731 Wheat 32s 10d per quarter.

Another very dry year, the driest of the series of four. Rainfall was about 66% of the average. It began with frost and snow, but summer, expecially September, was very hot and exceptionally dry.

The harvest again very good.

Jethro Tull published his classic, *The New Horse Houghing Husbandry*.

1732 Wheat 26s 8d per quarter.

The third of a spell of four dry years. Rainfall much less than normal.

An unseasonable cold spell occurred on May 1st and 2nd. On May 1st a very heavy snowfall at Edinburgh, and ice was thick enough to bear a man and horse. This late cold snap caused considerable floods when the thaw came, around May 7th, particularly in the Midlands.

The fatstock market at Smithfield is mentioned.

Mr Ellis published his book, *The Practical Farmer, or Herefordshire Husbandman*.

1733 Wheat 28s 4d per quarter.

Yet another dry year. The summer was exceptionally hot, especially in June and July. Autumn also fine and warm.

Heavy flooding in the North in March.

May was dry, with continuous north-east winds. Mornings and evenings were very cold, but day extremely hot. Tremendous thunderstorms broke the drought in the Midlands on June 26th and 27th, but dry weather resumed. 'So excessively hot for the most part of this month that it was scarcely tolerable; horses dying on the roads; much mischief by lightning' (Baker). Many cattle slaughtered because of scarcity of keep.

1734 Wheat 38s 10d per quarter.

The drought of the previous four years lasted till June 12th, 1734; but the second half of the year was exceptionally wet.

Great rains fell on July 13th, August 9th and 10th and October 1st.

Earthquake in Ireland did considerable damage, destroying 100 houses and 5 churches.

Flourishing trade in export of grain and malt. Wheat exports reached 498,195 quarters.

1735 Wheat 43s per quarter.

A rather wet year. The early part of the year was characterised by gales, rain and floods. A particularly violent gale on January 8th, when the barometer stood at 27.9. At Kilverstone, in Norfolk, a whirlwind on March 9th blew lead off the church roofs and blew water out of the river.

June and July were cold and wet, with few fine days. Corn was laid flat and hay spoiled. Heavy rains and consequent floods occurred on August 23rd, August 27th and September 7th. On the last named date floods were exceptionally severe at Droitwich and Coventry, with men, horses, sheep and bridges carried away.

On September 15th a hailstorm, which deposited hail two feet deep, destroyed most of the unharvested corn in Ayrshire.

An exceptionally severe thunderstorm at Ashby-de-la-Zouch on December 7th.

In January an epidemic of sheep rot is stated to have been the worst within living memory. William Ellis, of Little Gaddesden, Hertfordshire, comments that the carcases were a great nuisance in the highways. Thousands of sheep were not worth offering for sale, 100 being sold at Leighton market for sixpence each.

1736 Wheat 40s 4d per quarter.

Another wet year.

The winter was mild, with little frost but continuous rains. Heavy snow fell on February 8th and 21st. On February 16th the Thames rose to its greatest height for fifty years.

Almost continuous rain fell from the beginning of March to early July. 'All low meadows in the kingdom floated, and the hay and corn carried away or spoilt. The damage done almost incredible. In three days five inches of rain fell.'

A great storm occurred on October 9th, and a violent gale on November 12th. Exceptionally severe frosts in Edinburgh in mid-November. Gales, with much rain and flooding, continued throughout November and December.

1737 Wheat 38s per quarter.

Beginning of a dry spell which, with a few breaks, lasted for fourteen years.

The year started wet, with a remarkably heavy storm on the night of January 9th. Much flooding in the West Country, with many sheep drowned.

Similar violent thunderstorms occurred at Bristol on July 2nd, and in London on August 2nd.

Apart from these, the summer was dry, and June is recorded as a month of drought, though August was unusually cold.

Autumn brought heavy rains and gales, with floods.

Food riots occurred, directed against the export of grain. A load of wheat being taken from Salisbury to Southampton in May was stopped and smashed, the corn being scattered about the road.

1738 Wheat 35s 6d per quarter.

Written down as a dry year chiefly because of a drought in August and early September. The year started with gales and thunderstorms in January and February. April was cold and dry, but May and June brought wet weather and caused much mortality among sheep, through rot. July was in general a very dry month, but a very severe thunderstorm swept across the south Midlands on July 25th.

1739 Wheat 38s per quarter.

A rather wet year. There were violent gales in January and an exceptionally heavy thunderstorm on May 20th. Great damage to crops was caused by excessive rain and gales in September,

A notable feature was an unusually severe of frost which began on December 24th, but the worst effects were noted in the following year.

1740 Wheat 50s 8d per quarter.

A year of drought, the beginning of a spell of four exceptionally dry years.

The severe spell which started on December 24th, 1739, lasted for at least nine weeks, and cold weather continued till late in the spring. A Frost Fair was held on the frozen Thames from Christmas, 1739, till February 17th, 1740. Souvenirs were printed on printing-presses set up on the ice, and also on the ice of the Ouse at York.

'Many hens and ducks, even the cattle in the stalls, died of cold; the trees split asunder. Not only beer, but wine in cellars, froze. Deeply sunken wells were covered with impenetrable ice. Crows and other birds fell to the ground, frozen in their flight. No bread was eatable, for it was as cold and hard as a stone' (from Brocke's, *Contentment in God*). On the coldest night the temperature fell to 2°F.

The winter was also dry. Gilbert White records that at Selbourne 'not three hours continued rain from the beginning of November until the following April.'

The cold north-easterly winds continued throughout April and May, and snow fell in Yorkshire on April 22nd. Snow also fell in London on the night of May 16th/17th, and a severe frost occurred in north Yorkshire on May 30th. It was still cold in July, and the rainfall figures were redeemed largely by a very heavy thunderstorm in late July. Harvest was late and poor, and much fruit failed to ripen. Gilbert White records that fieldfares remained in England till June.

Autumn began early, a severe frost on October 12th giving ice half-an-inch thick in Kent. It was reckoned the coldest

October on record. A hurricane occurred in London on November 1st.

Jethro Tull died.

1741 Wheat 46s 8d per quarter.

Another hot, dry summer, following a hard winter. Many springs and ponds failed, but the harvest was abundant and gathered in good order. The drought was almost absolute from January to the beginning of June. Then, after a brief wet spell, it returned and lasted till the end of August. Autumn, with the exception of November, was fine and mild.

Agricultural prices were correspondingly low. Beef was sold for a penny a pound, and wheat for sixpence a stone.

A splendid year for partridges.

The sheep population of Great Britain was estimated at 16,640,000.

1742 Wheat 34s per quarter.

The third of the years of drought.

Heavy snow fell in northern England in January. A spring drought lasted well into April.

August was an unusually hot and dry month, though severe thunderstorms occurred in several places on August 18th. Pastures around Bristol severely damaged by swarms of grasshoppers at the end of August.

An excellent harvest.

1743 Wheat 24s 10d per quarter.

Another exceedingly dry year, with rainfall only 69% of normal. This degree of drought has never since been surpassed.

Drought was almost absolute in May, June, August and September, though a heavy thunderstorm swept through the west Midlands on August 18th. A fair amount of rain in July, when the unusual quantity of 5.23 inches fell in Rutland.

Harvest very good.

1744 Wheat 24s 10d per quarter.

The series of dry years was broken, a wet spell setting in during the third quarter of the year.

The year began with a severe winter and late spring, snow falling on April 2nd. April proved to be a wet month.

The summer and autumn rains caused much damage to crops in the fields, and, towards the end of the autumn, floods occurred, causing casualties among sheep and cattle.

1745 Wheat 27s 6d per quarter.

A wet, dismal year. Excessive rain in spring and summer; autumn cold and dry. A wasteful harvest, with much corn growing out.

The beginning of a severe epidemic of rinderpest among cattle, which lasted for at least ten years. This started a movement to plough up pastures and convert them to arable land, a trend which had the effect of depressing the price of grain.

About this date Robert Bakewell began his livestock breeding experiments at Dishley Grange, in Warwickshire.

1746 Wheat 39s per quarter.

A very hot, dry summer with heavy yields, particularly of barley.

The year began with extraordinarily mild weather, though severe frost occurred in the north in February, causing much mortality among cattle and sheep. After February a long, dry spell set in.

More outbreaks of rinderpest. First Act of Parliament for suppression of the disease and payment of compensation.

1747 Wheat 34s 10d per quarter.

On the whole, a long, dry summer, though with intervals of heavy rain.

A dry spell in early spring, till the end of May. Very heavy rain in June, which laid flat many fields of corn. Thereafter fine weather for harvest, which yielded well. Grass was scarce early in the year.

The wet spell around midsummer caused a serious outbreak of sheep rot, and rinderpest among cattle again caused much trouble.

Very dry autumn in Scotland.

An earthquake in the West Country on July 1st.

Sugar was first extracted from sugar beet (in Germany).

1748 Wheat 37s per quarter.

A changeable year, with moderately good harvests.

January was unusually cold, with much snow, occasioning loss of life among sheep. Another cold spell occurred in mid-February.

The summer was very hot, broken by occasional violent thunderstorms. In one at Crawfurd, Scotland, on June 27th, a flash of lightning killed 320 ewes.

A swarm of locusts reached England on August 4th, settling on vegetable crops.

Many gales in autumn.

1749 Wheat 37s per quarter.

A mild winter, followed by another hot summer. There were, however, some violent fluctuations in the weather, unusual numbers of thunderstorms occurring.

On May 24th and 25th 800-900 black cattle died in a violent snowstorm in Scotland. In another storm on June 3rd snow lay in Carlisle. In another on June 15th snow fell on Skiddaw; and on the 16th snow two inches deep fell at Stockport, with ice forming thick enough to hold a dog.

The harvest was excellent. Cattle plague continuing, the bill for compensation rose to £7,000 a month.

1750 Wheat 32s 6d per quarter.

Another very mild winter, followed by a hot summer.

Very warm from the very beginning of the year and the temperature climbed to 96°F on several days in July.

Earthquakes shook London on February 8th and March 8th, and Nottingham on August 23rd.

Several violent thunderstorms occurred during summer, including one at Gloucester on September 2nd, said to have given the heaviest rain ever known there. Last three months of year wet.

A magnificent harvest again.

William Marshall, visiting Devonshire, found not one wheeled carriage in the whole country, everything being carried by sledges or pack-horses.

1751 Wheat 38s 6d per quarter.

A changeable but, on the whole, wet year.

A spell of severe frost from January 26th to February 8th. March and May were excessively wet months, delaying sowing seriously. A thunderstorm with snow on May 6th. June was hot and thundery, and subsequently much rain made the harvest difficult.

1752 Wheat 41s 10d per quarter.

Mainly a wet year, with a disappointing harvest.

Blizzard with 27 inches of snow in mid-January. Very stormy in February and March.

Cool and damp for most of summer, and very heavy rains in August and September. 10,000 sheep reported destroyed by rain and floods in Wales on September 19th. Mild autumn till frost began on December 22nd.

1753 Wheat 44s 8d per quarter.

A dry summer with good harvest.

Much snow and frost in early part of year. Floods in February. Gales in March, April and May. Exceptionally heavy thunderstorm, with huge hailstones, in the West Country on June 2nd.

Rinderpest still raging.

1754 Wheat 34s 8d per quarter.

A late spring was followed by a fine summer and excellent harvest.

Violent thunderstorm, with huge hailstones, in Suffolk on July 28th.

Prices of stock and feedstuffs high early in year but low after harvest.

1755 Wheat 33s 10d per quarter.

A rather wet year, with a disappointing harvest. Exceptionally heavy rain in northern England in September.

1756 Wheat 45s 3d per quarter.

A very wet year, with poor harvest. Said to have been the wettest summer within living memory. Similar weather prevailed over most of Europe.

Hurricane did much damage in Cumberland on night of October 6th.

Prices rose alarmingly, and food riots occurred. Export of grain temporarily prohibited.

1757 Wheat 60s per quarter (thus nearly doubling its price in two years).

Reckoned a famine year, because of scarcity of grain from previous harvest.

At Cambridge a merchant was convicted of offering 6s a bushel for wheat for which only 5s 9d was asked. For this attempt to raise the price he had to forfeit £50 to the poor of Ely and £50 to the poor of Cambridge, besides apologising publicly on market day in Ely market place.

Prisoners in Somerset goals were allowed a daily ration of twopennyworth of bread, because of the high prices demanded.

The year was fine and warm, however, and a good harvest was reaped, to alleviate the situation.

Rinderpest plague now declining.

E. Lisle published his book, *Observations in Husbandry*.

1758 Wheat 50s per quarter.

An uneventful year, with good weather and satisfactory harvest.

1759 Wheat 39s 10d per quarter.

Another favourable year, with a hot, dry summer.

1760 Wheat 36s 6d per quarter.

The series of fine years continued. Summer dry and exceptionally hot.

Very violent storms in London on February 15th and north Wiltshire on February 16th.

Unusually mild autumn, with fruit trees, primroses and daisies in bloom.

Estimated to be 365,000 farmers and 200,000 agricultural day labourers in England and Wales.

About this date farmers in Norfolk started using oilcake for fattening bullocks.

William Cobbett born on March 7th.

1761 Wheat 30s 3d per quarter.

Yet another fine year, with a good harvest, following a mild winter. The dry spell ended, however, in late August and was followed by storms and a very wet autumn.

1762 Wheat 39s per quarter.

A year of great drought.

A blizzard in February lasted for 18 days, and snow in some places lay 10-12 feet deep. Nearly 50 people lost their lives, and many cattle died.

In the drought and heat a heavy harvest was gathered, but fodder became exceedingly scarce.

The drought broke in exceptionally heavy rainfall from October 26th onwards, especially in the eastern counties. 'Most of the cattle in the fields were carried off; likewise stacks of hay and wood, with the loss of all the hogs that were in their styes and yards, together with all the horses that were in stables; for the waters rose twelve feet in less than five hours, which was in the dead time of the morning, nothing of it being perceived at one o'clock; it reached the chamber windows before five, and the face of the water was covered with the bodies of the beasts that perished. The damage at West Ham alone was computed at £100,000. At Chelmsford, Cambridge and Norwich great damage was sustained, and at many other places; sundry persons perished, and several thousands of hogs, horses, horned cattle and sheep were drowned . . .' *London Magazine*

1763 Wheat 40s 9d per quarter.

Recorded as wettest summer since 1756, with much rain from July onwards. Harvest reasonably good.

1764 Wheat 46s 9d per quarter.

A stormy year but an excellent harvest. Very heavy thunderstorms in July and August. Grass abundant, and the hay crop good. Farmers in Hertfordshire were saying at the end of July that they were now expecting double the yield of barley that they had anticipated a month earlier.

After harvest prices fell accordingly.

1765 Wheat 48s per quarter.

A dry summer with heat waves. Harvest plentiful. Spring wet. A good year for all fruit and flowers.

Arthur Young published his, *Essay on the Management of Hops*.

1766 Wheat 43s 1d per quarter.

(*Note.* Until this date the prices given are for a nine-bushel quarter. Hereafter, and beginning with this year, they are for an eight-bushel quarter.)

A wet year, with a disappointing harvest.

The winter was severe, and heavy rains so interfered with haymaking that a party of haymakers brought their troubles and grievances to London and had a collection made for them on the steps of the Royal Exchange. Heavy flooding occurred, particularly in the eastern counties, 3,000 sheep being drowned at Great Upwell on July 30th.

Harvesting weather was fine and propitious, but the grain was light. As prices rose, so did discontent. A food riot occurred at Exeter, and at Hoxton an assembly of poor people seized grain from farmers and sold it at 5s 6d a bushel. The exportation of corn was prohibited.

Sheep rot took a heavy toll in autumn.

1767 Wheat 47s 4d per quarter.

A dry summer, with a moderately good harvest, Prices, following previous year's scarcity, were high till after harvest and occasioned much unrest.

Very severe frosts began in late December.

Francis Moore, a London linen-draper, invented a steam-engine for ploughing.

1768 Wheat 53s 9d per quarter.

A wet year with a poor harvest, the first of a series of such years.

The year began with a very cold spell, lasting till January 16th. Spring cold and wet. From June 7th onwards torrential rain fell, and rainy weather continued till the end of harvest. Both hay and grain harvests were disappointing. A whirlwind of remarkable intensity devastated a moorland valley near Tavistock on August 22nd. Fogs in September, and rain frequent for rest of autumn. Serious floods in November.

Food riots in many places.

Arthur Young published his book, *Tours of the Southern Counties*.

1769 Wheat 40s 7d per quarter.

Another wet year. One of the wettest on record.

First months of year wet but comparatively mild. March was dry and windy; April stormy. July brought warm, dry weather, but with August came more heavy rains. Rain and fine periods alternated throughout autumn.

Rinderpest struck again in autumn, and it was a bad year for sheep-rot.

1770 Wheat 43s 6d per quarter.

Yet another wet year, though the weather was reasonably fine for harvest.

The early part of the year was relatively mild, though March was frosty. April and May were cloudy and somewhat wet. June and July were mostly wet, to the detriment of the hay harvest. August and September were mostly fine, though with some storms, and from October onwards rain was almost continuous. Heavy flooding, particularly in autumn, on the Thames, Severn and the rivers of the eastern counties.

Wheeled farm vehicles introduced to Devon and Cornwall about this date.

Importation of China pigs on a large scale began.

Arthur Young published his book, *Tour of the North of England*.

Rinderpest outbreak stamped out by rigorous application of the slaughter policy.

1771 Wheat 47s 2d per quarter.

The excessive rain of the previous autumn was followed by frost. Little autumn wheat had been sown, and a backward spring gave small opportunity for making up the arrears.

Of this late spring Gilbert White records: 'Dr Johnson says that in 1771 the season was so severe in the Island of Skye that it is now remembered by the name of the Black Spring. The snow, which seldom lies at all, covered the ground for eight weeks, many cattle died, and those that survived were so emaciated that they did not require the male at the usual season. The case was just the same with us here in the south; never were so many barren cows known as in the spring following that dreadful period. Whole dairies missed being in calf together. At the end of March the face of the earth was naked to a surprising degree. Wheat hardly to be seen, and no signs of any grass; turnips all gone, and sheep in a starving way. All provisions rising in price. Farmers cannot sow for want of rain.'

However, warm rains in May altered the outlook, and summer proved dry and cool. The harvest was reasonably good, after all.

November was unusually warm and wet.

Arthur Young published his, *Tours of the East of England.*

1772 Wheat 50s 8d per quarter.

A disappointing harvest, in spite of fine weather.

The sowing season was very late, with prolonged periods of cold winds. From June to mid-August hot, dry weather prevailed, but the spell broke before much of the harvest was in. Autumn was mild and wet.

1773 Wheat 51s per quarter.

A disappointing harvest, in spite of what appears to have been a favourable year. January brought first frost and then rain; February, first frost and then showers. March was bright and sunny and April showery. June was sunny and showery, and July and August mostly dry. Much rain fell from September to November.

In spite of this apparently ideal weather pattern, wheat yields are said to have been very poor.

More food riots.

Turnips were very plentiful in autumn, and lean sheep consequently much in demand.

1774 Wheat 52s 8d per quarter.

A wet summer though hot. Harvest no more than average. Rainfall in winter and spring excessive. March floods in many places, from London to villages high on Salisbury Plain. A good year for grass, fruit and flowers.

A stormy autumn, with damage to shipping around the coasts.

Foundation of the Bath & West of England Agricultural Society.

This was the last of a long series of wet years, of which Gilbert White wrote: 'Land springs have never obtained more since the memory of man than during that period (for ten or eleven years prior to 1774) . . . Such a run of wet seasons, a century or two ago, would, I am persuaded, have occasioned a famine.'

1775 Wheat 48s 4d per quarter.

A dry, warm summer with an abundant harvest. Rain fell almost incessantly till the middle of March. The next three months were showery. A period of heavy rain occupied the first half of July, after which August was warm and mostly dry, though with some showers. The first three weeks of September were wet.

1776 Wheat 38s 2d per quarter.

Warm, dry weather after the first week of August gave good harvesting conditions, though, owing to inclement weather earlier in the year, the harvest was no better than average.

The winter of 1775/76 was unusually cold over most of Europe, including Britain. A blizzard occurred on January 7th, and snowfall continued for most of the month. A thaw came in February, but March brought much frost, and April and May were cold and dry. Gilbert White records that 'till 30th of May the fields were burnt up and naked, and the barley not half out of the ground, but by June 10th there is an agreeable prospect of plenty.' Autumn was quiet and still.

In this year Thomas Coke, afterwards Earl of Leicester, began experimenting with scientific farming on his estates at Holkham, in Norfolk.

1777 Wheat 45s 6d per quarter.

A fine, dry harvest, though of about average quantity.

After a hard winter, with frost and snow, much of March was unusually mild. Gilbert White records that March 26th and 27th were so hot that bees swarmed. The thermometer stood at 66°F in the shade. Heavy rains fell in May, June and July, but August and September were sunny. Mild weather lasted throughout the autumn.

1778 Wheat 42s per quarter.

Another warm, dry summer. Harvest excellent.

The winter was hard, with frost and some snow till the middle of March. The Thames was frozen over at Kingston. Frost and snow were also recorded in the latter half of April, and May and early June were cool. A very hot spell occurred in early July, and heavy storms fell in the second half of the month. Thereafter harvest was dry and sunny. Much rain in autumn.

1779 Wheat 33s 8d per quarter.

Yet another warm, dry summer, with an abundant harvest.

After a frosty January, the weather was warm and dry till the latter half of April. After a short wet spell, May was fine and sunny, and June brought considerable heat, with some storms. July and August were hot and dry, except for a stormy spell from July 18th to August 8th Autumn weather was changeable.

A violent storm on January 1st did immense damage all over the country. It was accompanied by a remarkable high tide off the Lincolnshire coast.

Some of the thunderstorms at the end of July were very heavy. Violent gales occurred around the middle of October.

Harvest was so plentiful that prices of all farm produce, including wool, fell to almost unprecedented levels, causing much distress among farmers.

1780 Wheat 35s 8d per quarter.

Although this proved to be another reasonably warm, dry summer, the crops were attacked by mildew, and the harvest therefore was light.

The first four months of the year were cold and grey, with the exception of March, which brought sunny, springlike weather. From May to the end of August spells of sunshine and rain alternated. Autumn was mostly fine, with a period of early frosts in November.

1781 Wheat 44s 8d per quarter.

Another very dry year, with hot weather in summer. Mildew again attacked the wheat crop, resulting in light yields.

'Spring was cold and late, though dry. Late April and most of May were mild and springlike, with light showers, and early June brought heavy rains. Thereafter dry weather prevailed till the end of October. Many wells failed and ponds went dry', says Gilbert White. It was the worst drought for over 40 years.

A cattle disease caused heavy mortality in the spring.

1782 Wheat 47s 10d per quarter.

By way of contrast with the preceding years, this was an unusually wet one. Harvest was poor and wasteful, though there was an abundance of grass.

Spring was late, with frost, wind and much cold rain till the beginning of May. Most of May and June were warm and dry, but during July and August it rained almost incessantly. A brief fine spell was interposed during the first fortnight of September, and then showery weather set in till the end of October. November and December were mostly frosty, November being recorded as the coldest on record to that date, with an average temperature of only 34.7°F.

50.26 inches of rain fell at Selborne, compared to 30.71 in the previous year.

Cattle and sheep were in poor condition, and food prices climbed rapidly.

Caterpillars destroyed thousands of acres of turnips in Norfolk.

1783 Wheat 52s 8d per quarter.

A hot though stormy summer. Crops reasonably good, though not a bumper harvest.

Winter was mostly wet, with strong gales, though occasional periods of frost. Spring was late, with cold, dry winds and a hard frost on May 5th. The summer proved oppressive and thundery, and September brought a longish period of heavy, driving rain. Thereafter, the autumn was mild and mostly dry.

Gilbert White records: 'This summer was an amazing and portentous one, and full of horrible phenomena: for, besides the alarming meteors and tremendous thunderstorms that affrighted and distressed the different counties of this kingdom, the peculiar haze or smoky fog that prevailed for many weeks in this island and in every part of Europe, and even beyond its limits, was a most extraordinary appearance, unlike anything known within the memory of man. From June 23rd to July 20th inclusive, during which period the wind varied to every quarter without making any alteration in the air. The sun at noon looked as blank as a clouded moon and shed a rust-coloured ferruginous light on the ground and floors of rooms, but was particularly lurid and blood-coloured at rising and setting. All the time the heat was so intense that butchers' meat could hardly be eaten on the day after it was killed, and the flies swarmed so in the lanes and hedges that they rendered the horses half frantic and riding irksome. All the while, Calabria and part of the isle of Sicily were torn and convulsed with earthquakes, and a volcano sprung out of the sea on the coast of Norway.'

In this year Arthur Young started his journal, *The Annals of Agriculture.*

1784 Wheat 48s 10d per quarter.

On the whole, a wet summer, with much damage to crops. Harvest about average.

The year began with a very severe spell, which lasted till February 19th. After a short, mild interlude, frost and snow returned, and spring was late. After May 12th the weather turned hot, with a wet spell at the end of June, but otherwise dry till July 18th. Then heavy rains, though with high temperatures, began and lasted till the end of August. Autumn was fine and mild, but an unusually cold spell occurred in December. From December 7th to 9th a blizzard raged, covering the ground with 12-15 inches of level snow. Severe frost followed, the temperature falling to 1° below zero F. 'The cold stripped holly trees of their leaves and killed furze and ivy' (Gilbert White).

The Highland Society was formed.

Cattle and sheep were in poor condition, and food prices climbed rapidly.

Caterpillars destroyed thousands of acres of turnips in Norfolk.

1785 Wheat 51s 10d per quarter.

A stormy summer in most places, though some districts had drought.

A mild beginning to the year, but at the end of January a a six-weeks' frost set in. The average temperature for March was 33.9° F, making it one of the coldest Marches on record. After further spells of frost, interspersed with mild breaks, spring proved late, with a drought in much of April and early May. The summer was in general warm, with frequent storms and showers. The same sort of weather lasted throughout most of the autumn, and harvest was consequently very late. Gilbert White records that at Selborne haymaking finished on November 9th and wheat harvest on November 14th.

Arthur Young states that hay and straw fodder of all kinds were scarcer and dearer than ever before known. The heavy storms of September 'beat great quantities of wheat which was then ripe out of the ears'. Some districts, however, had good hay crops.

1786 Wheat 38s 10d per quarter.

A warm summer with frequent showers. Harvest moderate to good.

The year started with a cold winter. Spring was mostly wet, with a short, sharp period of frost in early May. Thereafter much rain, with high temperatures. A pleasant autumn, with occasional short periods of frost. A longer period of frost and snow began on December 16th and lasted till the end of the year.

Much damage to turnips by both frosts and turnip-fly. Damage from turnip-fly alone was reckoned at £100,000 in Devon. Shortage of livestock food in spring.

The mangold was introduced into England from France.

A threshing-mill was invented in Scotland by Andrew Meikle.

1787 Wheat 41s 2d per quarter.

A fine summer with good harvest.

Changeable in winter and spring, with fair amounts of rain. June and July becoming increasingly hot, with thundery showers (but a hard frost in Hampshire on June 7th). August and September hot and dry, with autumn also very dry. Mild till the end of the year, with one short spell of hard frosts at end of November.

The autumn drought was such that the chalkland rivers of southern England, notably the Wylye, went dry in their upper courses.

W. Marshall published his book, *Rural Economics of Norfolk*.

1788 Wheat 45s per quarter.

A year of drought.

Winter and early spring brought changeable weather, with frequent showers. April and May were mostly dry and warm, with very little rain. A period of heavy rains occurred early in July, but thereafter the weather turned hot and dry again.

Accounts of the crops vary considerably from district to district. The chalk country of the south felt the worst effects of the spring and midsummer drought, with very little hay or pasture available in June. Baker records that 'upwards of 5,000 horned cattle perished from the dryness of the season'. The eastern counties, on the other hand, reaped a very good harvest.

During the short spell of heavy rains in early July, several violent storms swept the country. There are records of flooding in Scotland and the North, also in London. However, the worst storm just missed England; it came in from the Bay of Biscay on July 13th and swept right across France and the Netherlands. The hailstones are stated to have been as big as quart-bottles and to have taken three days to melt. Immense damage was done. Exceptionally heavy rain fell in East Anglia on June 28th.

On November 30th a spell of frost began which lasted until early January, 1789. The Thames was frozen over, and a frost fair held on it,—the first for many years. It is said to have been the coldest December on record, with an average temperature of 29°F.

William Marshall published his book, *Rural Economics of Yorkshire*.

1789 Wheat 51s 2d per quarter.

A late winter, backward spring and crops subsequently below average.

Frost and snow lasted till the third week of April, after which rain was frequent till the end of July. August was hot and dry, but showers were frequent again in September. The remainder of the autumn was unusually wet, which greatly interfered with the winter sowing.

The arctic spell which began on November 30th, 1788, lasted till January 13th. This severe weather caused a great scarcity of keep, as many acres of turnips were destroyed.

The wet weather was a contributory cause of a serious outbreak of sheep-rot.

Merino sheep were imported from Spain by George III.

William Marshall published his book, *Rural Economics of Gloucestershire*.

1790 Wheat 54s 9d per quarter.

A moderately good year for weather and crops.

A mild beginning to the year, though with short spells of

frost in January and February. A cold spell, with heavy snow, occurred in the middle of April. Summer was changeable, with warm dry spells alternating with cool wet ones. One dry spell, from August 6th to 24th, greatly assisted the harvest. Autumn was mostly wet, and there are records of floods (particularly in the North) and of gales. Much snow fell in mid-December.

Sheep-rot was again bad.

William Marshall published his book, *Rural Economics of the Midland Counties.*

1791 Wheat 48s 7d per quarter.

An excellent harvest, with good harvesting weather.

January was mild and wet; February brought a good deal of snow, with strong winds. March and April were dry but cold. May and June hot. July wet. July and August dry. October and November wet and stormy.

A remarkable fall in temperature occurred in southern England around June 12th to 15th. From 75°F the temperature dropped to 25°F, with accompanying snow and frost. The hay crop was good, but December frosts damaged the turnips.

The Veterinary College was founded in London.

1792 Wheat 43s per quarter.

Conflicting accounts of this year's harvest.

The *Gentleman's Magazine* records, 'Summer remarkably cold and ungenial all over England. It was uniformly wet, windy, cold and dark, excepting one dry week in August, when the heat was so excessive as to cause many deaths, and at the commencement of September all thoughts of summer were annihilated by severe frosts.'

Gloucester Notes and Queries records that 'spring and summer were wet and cold, hay and corn bad, wet winter, but neither frost nor snow.'

Baker states that 'summer and autumn were a continued series of wet weather. Both corn and hay greatly injured in harvesting.'

On the other hand, some places had a dry summer.

In general, however, the harvest was poor.

A bad year, too, for sheep-rot.

1793 Wheat 49s 3d per quarter.

A dry summer but no better than average harvest.

The temperature remained rather low for most of the summer, and rain in some districts came before the end of harvest and caused much damage. The hay crop was poor, and turnips did poorly in the drought.

The Board of Agriculture was first set up, with Arthur Young as its first secretary.

It is recorded that 116,448 cattle and 729,810 sheep passed through Smithfield Market this year.

1794 Wheat 52s 3d per quarter.

Another dry summer, but a deficient harvest. Wheat estimated to yield 14 bushels per acre, as against 20 in 1793. The low yield is partly accounted for by a drought in spring.

Another bad year for sheep-rot.

1795 Wheat 75s 2d per quarter.

A poor harvest.

The winter of 1794/95 was exceptionally severe. The arctic spell began in December, 1794, and January is said to have been the coldest on record. The temperature on January 25th was –6° F. and the average temperature for the month was only 23.9°F.

This cold spell lasted late, until about March 21st. Scotland had excessive snowfall, and in Europe the Zuider Zee was frozen over.

A late, cool spring followed, and the summer was also cool (more than 2° below the average temperature in July and August). There were frosts and cold east winds in June.

September had one brief spell of very hot weather, with the temperature climbing to over 90° on two days.

Reports on the harvest conflict. Most writers agree that it was below average in quantity (estimated at 12 bushels per acre), but the quality evidently varied. Some wheat is stated to have been good, but Baker records that in some the ears were not filled out and the capsules quite empty. Harvest was very early in some districts.

The cold spell in June caused many newly-shorn sheep to die.

Food was scarce, and food riots occurred.

The Speenhamland system was introduced, whereby farm

workers were entitled to a weekly sum, either from their own labours or out of the parish rates. The effect of this piece of well-intentioned legislation was to reduce most of the labouring population to the status of paupers. The system was worked out by magistrates and others at Speenhamland, near Newbury, Berkshire, on May 6th.

At Smithfield Market the average weight of beef cattle was 800 lb; of calves 148 lb; of sheep 80 lb; of lambs 50 lb.

1796 Wheat 78s 7d per quarter.

Crops no better than average; poor in the North.

A very warm January. The outstanding meteorological feature of the year was the intense frosts of December. On December 24th the temperature fell to −16°F, the lowest temperature on record.

William Marshall published his book, *Rural Economics of the West of England.*

1797 Wheat 53s 9d per quarter.

A very wet summer. Crops rather poor.

A backward spring, and summer rather cold till the middle of July. The temperature then climbed to 89° to 91°F for about ten days, breaking in tremendous thunderstorms, which did much damage. Thereafter the summer was wet and stormy.

1798 Wheat 51s 10d per quarter.

A fine, warm summer with reasonably good yields.

This followed a wet winter. Harvest was early, and harvesting weather fine. A tremendous storm occurred on September 12th.

Sheep-rot was prevalent at the end of winter.

The Smithfield Club was founded, and the first Smithfield Show held.

William Marshall published his book, *Rural Economics of the Southern Counties.*

1799 Wheat 69s per quarter.

A wet, unprofitable year.

Baker records: 'From its commencement to its close this season was, perhaps, as ungenial to the productions of the earth and to the animal creation as any upon record, and the inclemency extended over a great part of Europe. In this country, and particularly in the north of the island, many fields of corn were still uncut as late as November, and some were not cleared till the January following.'

About the beginning of August heavy rains, with cold easterly winds, reduced the summer fallows and turnip fields into a perfect mire, half rotted a great part of the hay, stopped the growth of the second share clover, laid down all the strong corn, and prevented the wheat from filling. The month of September was, on the whole, rather worse. Much of the wheat died at the root before the ear ripened, from excess of moisture. Oats suffered less, but they were nearly destroyed by two severe nights of frost on the 16th and 17th of October. Some beans were also injured.'

For once, all chroniclers tell the same story. From June 22nd to November 17th there were only eight days without rain. Harvest in many districts did not begin till September, and much of the corn was never got in.

1800 Wheat 113s 10d per quarter, (a consequence mostly of the previous wet summer).

Another disappointing harvest. Spring was late and cold, fine hot weather in July promised well, but before the harvest was well started rain set in again.

Baker records: 'A small part only of the crops was got in before heavy and almost incessant rains began. Of the wheat, that part which was secured in the southern districts of the island before the rains commenced, and therefore in good condition, proved to be light, coarse and unproductive. But the rains came on in August, and caught a considerable proportion in the fields, even in the south, and injured the whole of the crops in the north of the island. The crops were still worse in Scotland. In Mark Lane, in the course of the season, wheat reached the extravagant price of 184s per quarter.'

Baker also gives the following detail: 'October 30th, men were employed raking barley out of the lowlands of Chadenwych Farm, Mere (Wiltshire) and taking it to the higher ground to dry. It was carried to the rick November 6th . . .'

Several remarkable storms occurred during the rainly summer. On August 19th a thunderstorm swept across the Midlands, doing much damage. Hailstones eleven inches in circumference fell in Bedfordshire, killing hares and partridges. Another violent storm occurred on September 15th spreading over most of Britain.

Food correspondingly scarce, and more food riots occurred.

1801 Wheat 119s 6d per quarter.

A fine summer and an excellent harvest.

Winter was warm and wet, and spring cold, but an abundant harvest brought about a rapid fall in prices.

1802 Wheat 69s 10d per quarter.

Another fine summer, though yields of crops not quite as high as in 1801.

Spring was late; hard frosts occurred in June; and July was wet and cold. August and September were, however, sunny and warm.

A violent storm swept the north of England on August 18th.

William Cobbett started publishing his celebrated, *Weekly Political Register*.

1803 Wheat 58s 10d per quarter.

A rather cold year, with a moderate harvest.

April, May and June were all distinguished by cold, drying winds. Hoar frosts occurred in June, July and August. September was mild and wet, and the autumn and winter were also mild.

1804 Wheat 62s 3d per quarter.

A changeable year, with a moderately fine harvest but with cold winds and much chilly rain from May to July. Consequently the wheat crop was rather poor.

Violent storms in the West Country and the North on May 4th.

Imports of guano from Peru began.

1805 Wheat 89s 9d per quarter.

Another cold year, though with a warm spell in late August and September. Harvest average.

Violent thunderstorms in the London area occurred on May 28th and July 6th.

A late and cold spring.

John Lawrence published his book, *A General Treatise on Cattle*.

1806 Wheat 79s 1d per quarter.

A late, cold spring, followed by a hot summer. Temperature rose to 95°F in Essex in July. Violent thunderstorm on July 24th.

Crops average.

1807 Wheat 75s 4d per quarter.

Conflicting reports about this year's harvest.

In general it was good in the south, deficient in the north, July and August were fine and hot, enabling southern farmers to get most of their harvest gathered, but the autumn rains set in early. A hard frost on September 12th stripped many trees of their leaves. Much of the northern harvest was still in the fields, much damaged, in November. Winter frosts began on November 4th and continued till the end of the year, Scotland in particular suffered. The Irish potato crop, too, was a partial failure.

1808 Wheat 81s 4d per quarter.

A poor harvest on the whole, with much wet weather.

In winter and early spring much snow fell, and March and April remained cold. A heavy fall of snow occurred on April 15th.

July was extremely hot, a temperature of 99°F being recorded in Suffolk on July 13th. Heavy thunderstorms occurred on July 15th (St Swithin's Day), with hailstones 6 to 7 inches in circumference falling in Somerset. Much damage was done to crops, notably wheat, in these storms. Stormy weather followed, greatly hindering harvest.

Much trouble from sheep-rot.

A period of severe frost began in Scotland in December which lasted till January 26th, 1809.

1809 Wheat 97s 4d per quarter.

Another wet summer and poor harvest.

January was a month of snow and ice, with a particularly bad ice-storm on January 19th. Even birds were disabled by it, lying coated with ice on the ground in great numbers.

April was wet, but May fine and warm. Heavy rains began early in July and continued till October. Much corn sprouted, and wheat was affected by mildew. Floods occurred.

Another bad year for sheep-rot.

1810 Wheat 106s 5d per quarter.

Yet another wet, cold harvest, with poor yields.

Spring was late and cold, and much of the winter-sown wheat was destroyed by frosts. The hay crop was poor.

Fine weather in July and August helped the yields, though some districts had heavy storms. One such in the north did much damage and caused many casualties, on August 14th and 15th. October brought excessive rain, with severe floods on the Lincolnshire coast. Heavy rain continued in November.

Much wheat was affected by mildew, and the average yield was only 12 bushels per acre.

Much sheep-rot again.

1811 Wheat 95s 3d per quarter.

Yet another wet summer and bad harvest.

Much frost and snow in January; the Thames was almost frozen over in London, only a narrow channel being left open.

Spring was warm and sunny, but cold winds damaged the wheat at flowering time. July and August were cloudy and cool. Early September gave reasonable harvesting conditions, but the weather broke before the northern harvest could be got in. Much uncut corn was destroyed by violent storms in October, and the remainder was not harvested till the end of November.

Mildew and blight greatly reduced the yields of wheat. The pea crop was a failure, and beans also suffered. Barley yields were far below average, and oats were poor.

1812 Wheat 126s 6d per quarter, (the highest annual average recorded during the Napoleonic Wars)

Another wet year with a poor harvest.

An extremely bad fog in London in January. March brought prolonged frost, particularly in Scotland, with several blizzards in the middle of the month. Wet weather handicapped the hay harvest. July and August were cool and rainy; and winter began early. Corn not harvested before October was badly damaged by the weather.

Floods occurred in London in October.

1813 Wheat 109s 9d per quarter.

A fine, dry summer with an abundant harvest. General thanksgiving. The price of wheat in Portsmouth dropped from 123s 10d per quarter in January to 67s 10d in November.

A great frost began on December 27th, accompanied in London by a thick fog which lasted for eight days.

1814 Wheat 74s 4d per quarter.

An average year, as regards weather, but with yields rather below average.

The year began with an exceptionally hard winter. The fogs which began on December 27th dispersed on January 6th and were succeeded by snow and strong east winds. The snow fell for several days and lay on the ground for five weeks. Drifts were 15 feet high in places, and the snow is said to have been the deepest for forty years. A Frost Fair was held on the frozen Thames from February 1st to 4th. Temperatures fell to 13°F. A sudden thaw in mid-February was followed by severe floods.

Spring was late, and the hay crop light. Cool, showery weather lasted till the end of June, but July was very hot (temperatures up to 88°F). Between August 20th and September 20th there were no fewer than 25 days of brilliant weather, which greatly aided harvesting.

The barley harvest was good, but wheat was badly affected by mildew. Some barley, however, grew out in the swathe. Autumn was fine and allowed winter wheat to be sown in good order.

Heavy mortality among sheep in the cold weather. Sales early in the year saw sheep making low prices, and the ewes were in poor condition. In the autumn sales prices picked up, but enormous numbers appeared at the autumn fairs and many went home unsold. Several of the West Country fairs had the largest number of sheep ever entered. Wilton, for instance, had 63,000.

1815 Wheat 65s 7d per quarter.

A year of drought. An abundant corn harvest, but the hay crop was light, and turnips failed to thrive.

Fine weather from March till the autumn. Shortage of keep caused a glut of cattle on the markets.

Battle of Waterloo and end of the Napoleonic Wars.

As prices tumbled, farmers gave up their leases, and an agricultural depression began.

Sulphate of ammonia reputed to have been first manufactured in England.

1816 Wheat 78s 6d per quarter.

A wet summer with a very poor harvest.

A winter of storm, gales and floods. Spring was late and cold. Severe weather, with snow lying on ground, in mid-April.

Summer and autumn were also cold and wet, with very little sunshine. The temperature for July and August was 4.8° below average, and the heavy rain was accompanied by strong winds. Quite a heavy snowfall in the eastern counties on September 2nd, with severe frost in London and elsewhere.

Harvest began late, in many districts not before the end of August. Wheat sprouted and was in poor condition. In the Midlands and North much corn was still in the fields in November.

Sheep-rot was prevalent; hay scarce; and much livestock was sold for lack of keep.

One of the most disastrous harvests known.

Much distress, and food riots.

1817 Wheat 96s 11d per quarter.

Another wet summer with a poor, late harvest.

The year began with storms and floods. Spring was late. A phenomenally hot spell in the middle of June, with the temperature rising to 112°F at Weymouth on June 21st. Thereafter almost incessant rain till the end of August. The weather then cleared up and gave a fine though somewhat foggy September. Rain came again before all the harvest was in, and October was both wet and frosty. September proved to be the main harvest month, but by that time much of the corn had sprouted.

Another bad year for sheep-rot.

The first Hereford cattle were exported to America.

1818 Wheat 86s 3d per quarter.

A fine summer with an excellent harvest.

The year started with mild weather, but with chilly rain prevailing throughout much of the spring until the middle of May. The heaviest downpour occurred on May 8th, after which no rain fell in some parts of England, especially in the south, till early September. By September 3rd there had been a succession of 108 hot days, during which the temperature had averaged 65°F. July was particularly hot, the temperature rising to 121°F in the sun on July 23rd.

Harvest proved exceptionally early, some wheat being marketed on July 12th. In spite of shortness of straw, the yield was good and the quality excellent.

The hay harvest was meagre, but there was a luxuriant aftermath.

Mild weather continued throughout the autumn. Strawberries and rasberries were picked in Devon in early December, and spring flowers were in bloom.

Turnips suffered considerably from attacks of black caterpillars.

There was an abundance of food for cattle and sheep, but sheep-rot again caused serious losses.

In the autumn serious unemployment developed among the agricultural labourers of southern and eastern England, up to 60% being unemployed.

1819 Wheat 74s 6d per quarter.

An undistinguished year, with crops slightly below average in most districts, though variable.

Summer was hot and the fine weather prolonged. Spring-sown did much better than autumn-sown. Barley gave heavy yields of coarse grain, but wheat produced more straw than grain.

Snow on October 22nd.

1820 Wheat 67s 10d per quarter.

A warm, dry summer with a plentiful harvest.

Very severe winter, lasting till well into April. Great blizzards on January 9th and 15th. Temperature on the 15th fell to −10°F. Autumn-sown wheat much damaged by frost.

Great gales in early March, but later dry weather provided good conditions for spring sowing.

May was wet. June cool till the 24th, then very hot. July was thundery and cool, and August showery, but September provided excellent harvest weather.

Hay was rather scarce, and turnips did not thrive.

This year saw the origin of the famous Chevalier variety of barley, which was the most popular variety for the rest of the century.

1821 Wheat 86s 1d per quarter.

A wet, cold summer with a poor harvest.

Rain set in just before harvest and greatly hampered operations. The quantity of grain was better than might have been expected, but the quality suffered badly.

Snow fell in London on May 27th, the latest date ever recorded. Frosts occurred in June, which was an unusually cool month.

Autumn was distinguished by frequent gales and tempests. The weather remained mild, but there were widespread floods.

On September 1st a fall of *snails* occurred at Thornbury, Gloucestershire, covering six acres of land ankle deep!

William Cobbett published his, *Cottage Economy*. He began his famous, *Rural Rides*.

1822 Wheat 44s 7d per quarter.

A hot, dry summer with a very early but abundant harvest.

A remarkably mild January. Gales in February. May and June very hot. An early hay harvest of good quality. The corn harvest was also early, with wheat giving the best yields. Turnips were a poor crop, and potatoes average. Pastures dried up quickly, but fruit gave a bumper crop.

Sheep prices were much lower than in the previous year.

The first volume of Coates' Shorthorn Herd Book was published. Board of Agriculture wound up, as an economy measure.

1823 Wheat 53s 4d per quarter.

A wet summer and poor harvest.

The winter was cold and prolonged, with a great blizzard in north England on February 8th. Between June 29th and August 15th rain fell on 47 days, and every day but one in July was wet. Only once in the summer did the temperature rise to 76°F, and that was on June 1st.

The cool summer merged into a rainy autumn, with strong winds and gales developing. Severe floods occurred in southern England at the end of October.

A bad year for sheep-rot.

1824 Wheat 63s 11d per quarter.

Another wet summer, though warmer than the previous year. Crops a little below average.

The first three months of the year were exceptionally mild, and fine weather in June assisted a good hay harvest.

An exceptionally violent thunderstorm broke over London and the southern counties on July 14th, and thereafter rain was almost continuous. The rain was accompanied by heat (with a temperature of 86.5°F on September 2nd), which resulted in frequent thunderstorms.

Later in the year, gales were frequent. One of the worst storms on record hit the country on November 23rd, with great winds, pouring rain and terrific thunder, especially on the west coast.

The great German chemist, Justus Liebig, the founder of the science of agricultural chemistry, began his work at Giessen.

1825 Wheat 68s 6d per quarter.

A warm, dry year, with an early harvest rather below average.

The first few months were unusually mild, springlike weather prevailing throughout most of January. Spring was likewise dry, but cold, with sharp frosts. May was chilly, but June, July and August were exceptionally hot.

The June heat-wave resulted in an early hay-making, of good quality though rather below average in quantity. Grass dried up quickly, and pasture became scarce. In July ponds and streams dried up, and the heat was stifling.

Baker records: 'The young rooks entered our gardens, as in severe frosts, with open bills, panting. Horses dropped on the roads. The leaves of apple and filbert trees withered up. Gooseberries hung shrivelled on the leafless bushes, and potatoes sold in Bristol for 24s the sack. Bean crop defective, ripe by August 1st. Water scarce. Cows nearly lost their milk; and from July 18th to 24th butter could not be made to harden. Early in August rains fell, and continued seasonably until September, and by the end of the month the profusion of feed in the pastures was astonishing.'

1826 Wheat 58s 8d per quarter.

Another year of drought, though harvest seems to have been about average.

January was frosty, and spring cold and dry. The summer continued dry, though hot. Harvest was extremely early, and some of the corn had such short straw that it had to be pulled instead of cut. Some spring-sown corn had no rain from the time it was sown till it was reaped.

Plagues of caterpillars on hedges and fruit trees in spring; and American blight was abundant.

1827 Wheat 58s 6d per quarter.

A fine summer with an excellent harvest.

The year began with snow and frost, but spring came early, and the summer was uneventful. Early in the year cattle and sheep were cheap, through scarcity of fodder from the previous year's drought.

Autumn was rough and windy.

1828 Wheat 60s 5d per quarter.

A wet summer with a disappointing harvest.

Winter was mild, and the summer rains were concentrated chiefly in the period July 6th to August 15th. The rains of early July were exceptionally violent, causing considerable flooding. Hay was washed away and corn-fields laid flat.

A very bad year for sheep-rot.

1829 Wheat 66s 3d per quarter.

Another very wet summer, with a poor harvest.

Only four fine days between June 16th and September 20th. Rather cool, though not excessively cold.

Scotland had a severe drought in May, June and July, followed by excessive rain in August, with disastrous floods and much damage.

Considerable snowfall on October 6th and 7th.

Heavy mortality among sheep and cattle.

1830 Wheat 64s 3d per quarter.

A changeable summer. Accounts of the harvest vary, some recording a good harvest and some estimating it as less than average.

Blizzards in mid-January and severe frosts in February. March was fine, warm and excellent for sowing. July was very hot, but August and September were wet.

Sheep-rot very prevalent. In the winter of 1830/31 two million sheep are said to have died of it.

Machinery-smashing riots all over southern and eastern England, in protest against unemployment and the introduction of the threshing-machine.

About this date nitrate of soda began to be imported from Chile and Peru.

1831 Wheat 66s 4d per quarter.

Another wet summer, but with crops about average.

June was dry, but July and August, which were warm, brought much rain. A wet autumn.

An unusually severe frost, with ice nearly half-an-inch thick on ponds, occurred on May 6th and did much damage to fruit blossom and to grass.

More sheep-rot casualties.

1832 Wheat 58s 8d per quarter.

A rather wet summer, with crops about average.

A dry spring. Wet in June and August; relatively fine in July and September. Two heat-waves, one in mid-August and one at the end of September. A wet autumn.

1833 Wheat 52s 11d per quarter.

A fine, warm summer and plentiful harvest.

Winter and spring were mild and wet. May and June, fine July, stormy. August and September, dry. The hottest spell occurred in May.

In this year the first practical steam plough was produced.

The Royal Jersey Agricultural and Horticultural Society formed.

1834 Wheat 46s 2d per quarter.

A warm, dry summer and a bumper harvest.

A late, dry spring, and a hot, dry summer, except for July, when over 7 inches of rain fell. Much of the July rain fell in heavy thunderstorms.

The case of the Tolpuddle Martyrs.

A new Poor Law Act passed.

1835 Wheat 39s 4d per quarter.

Another favourable summer with an excellent harvest.

After a late spring, with a snowfall on April 16th and 17th, early summer was fine and sunny. Heavy rain fell in late June and early July, after which another fine spell set in and lasted through harvest. Much straw as well as grain.

William Cobbett died.

1836 Wheat 48s 6d per quarter.

A backward harvest, but rather above average. Poor in Scotland.

The 1835/36 winter was unusually severe, and spring was rather wet. Summer was hot in England (though cold in Scotland), with fairly frequent rain. Autumn was also wet, and heavy snow which fell on October 29th remained on the ground for a week. A very violent gale occurred on November 29th.

An unprecedented blizzard swept the country on December 25th and succeeding days. Drifts up to 50 feet deep covered the whole country and disrupted communications and all business for a week or more.

Black caterpillars took heavy toll of the turnips, making keep scarce.

The combine-harvester was patented in the U.S.A.

The Tithe Commutation Act passed.

1837 Wheat 55s 10d per quarter.

A rather warm, dry summer, but harvest below average.

A cold, late spring, with severe frost and snow in April and up to May 23rd. Fodder consequently very scarce. Dry in July and September, but wet in August. A fine autumn.

Thomas Coke of Holkham created Earl of Leicester.

The agricultural settlement of New Zealand began.

1838 Wheat 64s 7d per quarter.

A cold, wet summer and poor harvest.

The year began with a period of frost which lasted for nearly two months. A temperature of −14°F at Beckenham was the lowest recorded in the London area in the nineteenth century. Many evergreen shrubs were killed.

Spring dry and cold. May fine and warm, with the exception of a very cold spell in the middle of the month. Severe thunderstorms in June. July and August cool, but September fine and warm.

Fodder was scarce and dear in spring, and grass came late. Much winter wheat had to be re-sown, and the wheat crop was light. Barley quite good.

Foundation of the English Agricultural Society—later the Agricultural Society of England.

The Jersey Herd Book began.

1839 Wheat 70s 8d per quarter.

Another wet summer with poor yields.

Spring was late, with much snow in the middle of May. June, moderately fine, but rain fell almost incessantly in July, August and September. Much damage to crops. November was also unusually wet.

An exceptionally severe gale occurred on January 6th, blowing down houses and wrecking ships. On February 21st and 23rd the phenomenon of 'red rain' occurred in southern England. This was due to enormous quantities of dust, estimated at 10 million tons, carried from the Sahara.

The first Royal Show was held at Oxford.

Foot-and-mouth disease first appeared in Britain, causing a serious epidemic.

1840 Wheat 66s 4d per quarter.

A harvest rather above average.

The winter was mostly dry and frosty, with easterly winds in March providing a good seedtime. May and June were showery and rather cool, and July was cold, August began sunny and warm, but the fine spell was interrupted in the middle of the month by severe storms of wind and rain. The fine weather was back in September for the main harvest, and this continued into October.

The hay cut was light but of good quality. A bumper fruit year.

Much sheep-rot in places, and a severe epidemic of foot-and-mouth disease. Also of pleuro-pneumonia.

The German scientist, Justus von Liebig, published his book entitled *Organic Chemistry in its Applications to Agriculture and Physiology*, showing for the first time that plants fed on certain chemicals to be found in the soil and in the air.

1841 Wheat 64s 4d per quarter.

A wet, late season, but harvest about average.

January and February, cold with much snow. A sudden thaw of deep snow on January 14th caused severe flooding, even in normally high and dry country, such as Salisbury Plain; March, April and May, warm, with some showers. A cold spell in early June, with ice on ponds. Then stormy weather for rest of June, all July and first half of August. Some fine weather in late August and early September, but autumn very wet. Summer cool.

Little autumn sowing was done.

A severe epidemic of foot-and-mouth disease.

Contagious pleuro-pneumonia of cattle made its first appearance in England.

1842 Wheat 57s 3d per quarter.

A fine, dry summer, with good harvest.

Cold in January, and a cold, dry spring. On June 19th the first rain in two months fell. The hay cut was consequently light. July and August were dry, and August very hot, but the fine weather was interrupted by some violent thunderstorms, particularly in the eastern counties. Snow fell on October 26th.

The Farmers' Club was founded.

Patents for the production of superphosphate were granted to Lawes in England and to James Murray in Ireland on the same day, May 23rd.

1843 Wheat 50s 1d per quarter.

A mostly fine year, with a plentiful harvest.

January was mild and stormy, but February and March were mostly cold, with much snow. Bright, warm spells in March and April, but May was very wet. June was first showery and then fine, and July very hot, with the temperature reaching 90°F. August was likewise hot, but punctuated by several tremendous thunderstorms. September was unusually hot, the temperature reaching 85°F. Autumn frosts began on October 13th.

The hay crop, like the corn harvest, was excellent.

The Rothamsted Experimental Station was founded by Sir John Bennet Lawes.

Farmer & Stockbreeder first published.

1844 Wheat 51s 3d per quarter.

A year of drought with disastrous harvest.

After wintry weather in January and February, a drought began on April 14th and lasted till June 24th. It was broken on June 25th by a great storm, and sultry, showery weather prevailed until the middle of August, when another hot spell began. September and October were very warm, with thunderstorms.

Much of the spring corn failed to germinate till after the midsummer rains. This was consequently not harvested till October or November, and some of it never ripened. Autumn-sown corn yielded quite well, but spring corn was virtually a failure. The hay cut also was very meagre, as were all fodder crops.

1845 Wheat 50s 10d per quarter.

A cold, late season, with a fairly poor harvest.

An arctic spell which began on January 27th lasted till March 21st, with heavy snowfalls. The coldest February for fifty years. April was warm, but May cold and dull. After a fine June, July brought thunderstorms, and August was dull, cloudy and wet. September provided a reasonably fine month for the harvest. Autumn mild and fairly dry.

In consequence of the poor hay harvest and fodder crops of the previous year, keep was very scarce in the first half of 1845, and the late spring aggravated the situation. Sheep and cattle were in poor condition.

Severe epidemic of foot-and-mouth disease in autumn.

The first year of potato blight.

The Royal Agricultural College was founded at Cirencester.

1846 Wheat 54s 8d per quarter.

A fine, dry summer and plentiful harvest, at least, of wheat. Barley was below average.

The year began mild, and spring was unusually early. May was dry, and from June to September dry, hot weather prevailed. It was stated to be the hottest June on record, with the temperature rising to 95°F, and the average for the month being 65.3°F. In Ireland a 25-day period of great heat and drought came to an end on June 18th. August brought severe thunderstorms to some districts, particularly to London on August 1st.

The hot, sultry weather of June and July was contributory

cause of the devastating onslaught of potato blight, particularly in Ireland. This was the first year of the Irish potato famine.

Severe gales in autumn.

Keep very plentiful in spring and early summer.

Prohibitions removed on the importation of cattle from overseas.

First Hereford Herd Book published.

Repeal of the Corn Laws.

1847 Wheat 69s 9d per quarter.

Another hot dry summer with a plentiful wheat harvest, though of low quality.

A very dry and cold winter. Less than an inch of rain per month fell in January, March, April and July, and only 1.01 inches in April. July and August were unusually hot. At Greenwich the rainfall for the whole year was only 17.8 inches.

Wheat suffered much from mildew. Oats gave an average crop but barley a very good one. Owing to blight, the potato crop was a failure, and famine intensified in Ireland. Much mortality among sheep in early spring.

A bad outbreak of sheep pox, starting in sheep imported from Germany.

The coprolite deposits of Suffolk, Cambridgeshire and Bedfordshire were recognised as a valuable source of phosphates, and mining began.

1848 Wheat 50s 6d per quarter.

A wet harvest, with yields below average.

A cold and frosty January; then mild, wet weather for three months. May was cold and dry; June showery; July fine and hot; and August excessively wet. September provided a brief break for harvesting, and rain fell in abundance again in October.

Much wheat was badly sprouted; barley was variable; oats a poor crop.

Large White pigs were first exhibited at a Royal Show (held at York).

1849 Wheat 44s 3d per quarter.

A mild year, with an excellent harvest.

Rather hard weather in the early part of the year. An extraordinarily severe storm occurred in northern Scotland from January 24th to 26th, causing great damage at Inverness. A great blizzard raged over southern England on April 19th.

Thereafter a fine, warm summer, August being particularly dry. All crops good.

An Act of Parliament was passed enabling landowners to borrow money on the security of their estates for the purpose of financing long-term improvements of their property.

1850 Wheat 40s 3d per quarter.

A wet, cold season with a poor harvest.

A very severe January. March unusually cold and dry. June, August and September were rather dry, but July stormy. The weather remained cool throughout. In a week of unusually cold weather in Scotland around August 22nd, the snow lay low on the Cairngorms. Much damage was caused to the harvest by high winds and by mildew.

The first volume of the Devon Herd Book was published.

Buttermaking churns worked by horses were in use.

1851 Wheat 38s 6d per quarter.

A late harvest but rather above average.

A mild, wet winter. Dry in May and very hot in June, with temperature of 91°F at Chiswick. Autumn dry.

1852 Wheat 40s 9d per quarter.

A good harvest spoiled by heavy rain.

January wet, but after that a very dry spring till May 17th Much rain till July 4th, then very hot till beginning of August. Temperature reached 97°F at Chiswick.

August excessively wet, and rains continuing through September and October. The wind and rain, accompanied by high temperatures, caused much damage to ripening grain and produced a bad attack of mildew on wheat.

The September rains caused widespread flooding, especially in the Severn valley. The Journal of the Royal Agricultural Society states that 'all the vale of the Severn was one wide spreading sea. Its surface was covered with uprooted trees, crops and drowned animals.'

The autumn was too wet for much ploughing or sowing.

Severe epidemic of foot-and-mouth disease.

1853 Wheat 53s 3d per quarter.

Another very wet summer, with a poor harvest.

Much rain in April, and again in June, July, August, September and October. July rainfall was over 6 inches. The summer was in general cool as well as wet.

The Journal of the Royal Agricultural Society reports: 'Very disastrous summer floods. Hundreds of acres of meadow in low lying districts were cleared of their hay, and corn crops were flattened and spoiled. Many sheep swept away. Immense damage done to sheep, hay, and corn in Essex.'

Baker says: 'Bad crops of corn. Hay made badly. Wheat very thin plant owing to wet autumn last year. Rot in sheep. Cold season and very wet. Flakes of snow fell in June.'

Sheep-rot prevalent.

The Field first published.

1854 Wheat 72s 5d per quarter.

A fine summer with a good harvest.

The New Year opened with the countryside deep in snow, which melted on January 7th. Rest of January and February rather mild. March and April dry. Cool in May, June and July. Warm, sunny weather in much of August and September. October and November, dry and mild—ideal for wheat sowing.

Baker says: the wheat crop was the heaviest for many years, though it was late. The early hay crop was light, but some was not made till August. Oats and barley also gave good yields.

Rise in corn prices due to cessation of imports of Russian wheat, owing to outbreak of Crimean War.

1855 Wheat 74s 8d per quarter.

A moderately good summer, but crops rather below average, through rain damage in July.

A very hard winter, with much snow. Reputed to have been the coldest February on record, with the temperature falling to 1°F on the 18th. The average temperature for the month was 29.4°F. A cricket match was played on the frozen Thames at Henley on February 12th.

Cold, north-east winds prevailed till May 10th. June was warm and sunny, as were the first ten days of July. August and September provided good harvest weather till September

13th. Much rain fell in October, with severe storms. November dry and fine—ideal for wheat sowing.

Turnips, swedes and rape much damaged by turnip fly in June. The wheat crop, which had been damaged by the winter frosts, was badly knocked about by heavy rains in the last week of July.

1856 Wheat 69s 2d per quarter.

A warm, dry summer with variable crops, though good on the whole.

A dry, cold spring. Excellent hay-making weather in June and July. Harvest weather good in early August, but rain later did much damage. Early crops, notably of wheat, therefore yielded well, but barley was spoilt in many districts. September provided good harvest weather again, so the later crops of barley gave reasonable yields.

1857 Wheat 56s 4d per quarter.

A fine summer and good harvest.

A mild and wet spring. Good haymaking weather in June, and an excellent hay harvest. Exceptionally hot, sunny weather from July 10th to late August. Temperature rose to 92.5°F at Greenwich. Occasional heavy thunderstorms.

A fine, dry, warm autumn, excellent for sowing. Roses, fuchsias and many other flowers were in bloom on Christmas Day.

1858 Wheat 44s 2d per quarter.

A very dry summer, with good crops.

Spring cold and late. June hot and dry,—good haymaking. July stormy. Good harvesting weather in August and September. The temperature climbed to 97°F at Chiswick.

The wheat crop in most districts was abundant; oats and barley not so good. Turnip fly were active early in the season, but a good crop was obtained. Many ponds and wells dried up in summer.

1859 Wheat 43s 9d per quarter.

Yet another warm, dry summer with good crops.

Winter very mild, and spring early and warm. June, July and August were unusually hot, the temperature rising to 93°F at Greenwich in July. The average temperature for that month was 68.1°F.

Haymaking and harvest both favourable. Harvest very early, much being gathered in July. Some wheat crops injured by ripening too rapidly. Barley also good.

On October 25th a violent storm swept the country, causing much loss, particularly in shipping. This led to the introduction of gale warnings by the Meteorological Office.

November was fine and dry, allowing much autumn sowing to be done. December brought severe frost and snow, with a blizzard at the end of the month.

The first practicable traction engine was introduced.

1860 Wheat 53s 3d per quarter.

A cold wet year with a very deficient harvest.

Spring wet, cold and late. Much mortality among sheep. Fodder scarce.

June entirely wet. Fine spell for haymaking in early July. Wet again throughout second half of July and much of August.

Baker writes: 'Very little corn cut till September, and harvest not finished till November. Some corn harvested tolerably well beginning of September. Very deficient crop of wheat, and the quality very bad. The temperature was so low that there was little if any sprouted corn. Wheat was harvested in September, but scarcely any barley carried till October. Barley not bad quality, but condition bad, and some of it stained. Roots very bad; plants good but bulbs small. Great sheep-rot.'

The year was wet everywhere except in the north-west of Scotland. Heavy snow fell in the northern Pennines on May 27th.

On Christmas Day the temperature fell to 15°F, making this one of the coldest Christmasses on record.

The first reaping-machine was introduced.

1861 Wheat 55s 4d per quarter.

A moderately good summer, but crops were light.

Winter was severe, Though February and March were mild, east winds prevailed during April and May, making a late spring. Very severe gale occurred on February 21st, destroying part of the Crystal Palace. Another on May 28th destroyed much shipping.

June was warm and showery. The early hay crop was good,

but some of the later was spoilt by rain. August and September were mostly warm and dry.

Wheat rather light, but of excellent quality. Oats good. good. Barley poor. Very little fruit.

1862 Wheat 55s 5d per quarter.

A cold, backward summer, with crops below average.

Much rain in March and April interfered with spring sowing. May and June also wet. July and August cool and stormy. September a useful harvesting month, but October very wet.

The wheat crop gave particularly low yields, though the quality was good. Oats average; barley, very poor quality.

1863 Wheat 44s 9d per quarter.

A fine summer and abundant harvest.

A rather dry spring. Much rain early in June, but remarkably fine weather from June 20th to August 7th. Autumn mild but with many gales.

The hay was ready early and some of the earliest was damaged by rain, but in general the crop was heavy. Wheat was exceptionally good—the best harvest for over 30 years. Barley and oats average.

In a storm at Edinburgh on September 23rd 2.4 inches of rain fell in two hours.

1864 Wheat 40s 2d per quarter.

A year of drought, with moderate crops. A fair amount of rain in spring, but a drought which began in June continued into September. Occasional storms relieved it in places, but not everywhere. The temperatures were also below normal.

Baker records: 'Very dry summer. Some districts suffered greatly from drought, and in general the turnip crop was a complete failure, but July 17th we were favoured with a very heavy thunderstorm, which filled our ponds and kept the turnips growing. Our swedes were never better. Mangolds were small, and the late turnips suffered greatly from grub. Crop of wheat good, but not so bulky as last year; quality excellent. Barley crop good, and the corn heavy. Oats light in grain, but an average crop. Beans and peas almost a failure.'

The hay cut was light though of good quality.

1865 Wheat 41s per quarter.

Another year of drought, with light crops.

First months of the year rather wet. April was hot and so was September, both being reckoned the hottest known. In between, some districts had their rainfall brought up to near average by storms, but others had very little rain. October was excessively wet.

Baker records: 'One of the longest summers on record. Hot, bright weather set in in April, almost immediately succeeding sharp frosts, which continued till close on Lady Day. Wheats generally looked like a fallow till the end of March, then they shot off at a wonderful rate. June was scorching; nearly everything dried up, and many days scarcely a breath of air. Light crop of hay. In this district, a good yield of wheat, varying from 7 to 10 sacks per acre. Rather a catching wheat harvest. Early wheats harvested well, then a week of wet weather. Much wheat still out, and most of the barley, much of which went together damp. After this, a second summer, and beginning of October turnips were almost dried up. Swedes much mildewed. Rape like brown paper. October 10th, rain set in and set everything growing. No frost to stop ploughing one day to end of the year. Sheep suddenly rose to an extravagant price the latter end of August, in consequence of an outbreak of cattle plague in the London dairies . . .'

The outbreak of cattle plague to which Baker refers was an epidemic of rinderpest, which caused the death of over 253,000 animals, chiefly cattle.

1866 Wheat 49s 11d per quarter.

A rather wet year, with crops a little below average.

Baker records: 'January was a very wet month, with little frost. The 11th was one of the roughest days on record. Snow drifted tremendously, and quite blinded those exposed to it. Great loss of life on our coasts. Many sheep buried in the snow. February was wet, except last week, when beans were planted. March was wet and windy. Sowing done 12th to 15th, then stopped till 26th . . . April was stormy, and catching time for barley sowing till latter end, when it was fine and warm. May, cold easterly winds. Cut watermead for hay on 21st; carried it on 29th. July was stormy for the first week, after which meadow hay made well. August brought unsettled weather, but the wheat, oats and some barley put together sound, but not thoroughly dry. September was wet from beginning to end. Barley all black but not sprouted. Began barley cutting on August 20th; finished October 10th. October set in foggy the first week, then easterly winds, and the barley not harvested got thoroughly dry. Wheat sowing done well in this month and November . . .'

Apart from June and July, this was a cool summer. Heavy thunderstorms around June 27th.

Snow on the Yorkshire Wolds on May 1st.

The rinderpest epidemic abated.

Cattle Diseases Prevention Act was rushed through Parliament in only two weeks in February and was largely responsible for overcoming the disease.

1867 Wheat 64s 5d per quarter.

A rather wet year, with deficient crops.

After much rain in January and February, the latter part of February was fine and dry. March was exceptionally windy and cold. with some heavy snowfalls. April and May were rather cold and backward. The first week of June was stormy, but from June 8th to July 1st only 0.08 inches of rain fell.

On July 26th the Royal Observatory recorded the heaviest rainfall ever measued in twenty-four hours in England—3.67 inches.

August, September and October were unsettled; and November fine and rather mild. Heavy thunderstorms occurred in southern England on August 19th, after a very hot day.

There was a poor lamb crop, owing to bad weather in March. Seeds hay was quite good but meadow hay much damaged. Wheat was variable; oats about average; barley average for weight but poor in quality

First consignments of tinned meat imported from Australia.

1868 Wheat 63s 9d per quarter.

A year of drought, but with good crops.

January was a stormy, wild month. Gales and heavy rain caused severe flooding in Scotland towards the end of the month. February and March were mostly fine and dry, though a few wet spells.

The first really hot weather was encountered in May, the temperature rising to 90°F at Tonbridge on May 19th.

Thunderstorms followed, but June brought excessive heat and drought. This continued into July, the temperature rising to 100.5°F on July 22nd. There were some storms, but the heat returned. Particularly violent thunderstorms raged on August 11th and 12th and did much damage to crops, but the temperature remained very high throughout the month and well into September. The temperature was 92°F on September 7th.

Autumn rather cold, and winter stormy but mild.

All corn crops were good, wheat being well above average. Pastures were burned up and most fodder crops damaged or killed. Shortage of keep resulted in many dispersal sales or cattle and sheep at low prices.

1869 Wheat 48s 2d per quarter.

A dull cloudy year, with crops rather below average.

January and February were mostly mild and wet. March was mostly fine and cold, but with a heavy snowstorm, accompanied by thunder, on the 27th. April warm and sunny. May cold and wet. Hard frost in early June, with snowfall in the north. Hills in Yorkshire and Westmoreland were covered with snow in the middle of the month. July fine and warm. August mostly fine, though with some rain. Good weather lasted till September 10th, when a heavy thunderstorm caused much damage. Thereafter stormy till October, when another spell of fine weather enabled harvest to be completed. November fine and fairly mild.

Hay was good. Wheat rather poor. Much mildew. A moderate crop of turnips.

Severe epidemic of foot-and-mouth disease.

1870 Wheat 46s 11d per quarter.

A rather dry year, with excellent crops.

A rather cold spring, with little rain. Keep became short. May also was dry, cold to start with and then hot. The summer was hot and dry, with occasional violent thunderstorms which missed some districts entirely. One such storm with monster hailstorms fell in north Dorset on June 16th; another in Yorkshire on July 9th. Harvest weather was good, and the autumn fine and mild, but November and December became unusually cold.

Crop results very variable, according to the distribution of the storms. Wheat was, on the whole, excellent, and barley and oats good. Hay was of good quality but deficient in quantity. Some farms had good root crops, others reported failures.

Another severe epidemic of foot-and-mouth disease.

1871 Wheat 56s 8d per quarter.

A rather wet year, with crops about average but somewhat damaged.

The ten years from 1871 to 1880 were unusually wet, the average rainfall being about 20% above normal.

A wet spring on the whole, though with some dry spells. May dry and cold, but June very unsettled, as was July also. August provided a good harvest month, and the fine weather continued into September. October and November were cold and dry. The weather varied considerably from place to place, many districts recording heavy rainfall in late September. Cirencester had 6.70 inches during the month.

Early hay was good, but the later crops were much damaged by rain. The wheat yields were average but badly laid and much affected by mildew. Barley crops up to average for weight but not for quality.

Rinderpest was completetly eradicated in Britain.

1872 Wheat 57s per quarter.

Another wet summer and poor harvest.

Wet and fine spells alternated in spring, but much barley was not sown till May. May was cold and wet, 'the wettest May for many years'. Snow fell on April 21st in southern England, and a hard frost occurred on May 12th.

June and July had more than average rainfall, but mostly in heavy storms, and there were fine intervals for haymaking. August and September were relatively fine, but autumn was excessively wet. At Marlborough, October, November and December each recorded more than 5 inches of rain. A violent gale occurred on December 8th.

Many districts made good hay. The wheat harvest was gathered during fine weather in August but had been much damaged by the July storms.

Pleuro-pneumonia of cattle caused high mortality.

Formation of the National Agricultural Labourers' Union.

1873 Wheat 58s 8d per quarter.

A showery summer with a poor harvest.

January wet, with deep snow on the 30th. Heavy snowstorms in February. Spring cold and backward, lasting well into summer. June cold and dry. July, sunshine and showers. August and early September, stormy. Second half of September fine. Autumn fairly fine and open.

A smoke fog in London from December 7th to 13th caused many deaths.

Crops much affected by the poor seedtime and by frosts when the plants were in bloom. Low yields of wheat.

At this date half the land in England and Wales was owned by as few as 2,250 proprietors, in estates averaging 7,300 acres each.

1874 Wheat 55s 9d per quarter.

A year of good harvests.

January and February were wet and mild. March, dry—a good sowing month. April, showery. May, dry and cold. June, July and August were fairly dry. Some heavy thunderstorms fell about midsummer. Rain set in during September, and the rest of the year was wet, October unusually so.

Wheat excellent; barley rather light but of good quality, except that some was damaged by rain; oats, very poor; hay, good but rather light. Roots did badly during the summer drought.

First shipment to Britain of refrigerated meat—this from the U.S.A.

The British Beekeepers' Society formed.

1875 Wheat 45s 2d per quarter.

Very wet summer, particularly in July; crops suffered accordingly.

A wet January, followed by a fairly dry spring. Much rain in June even more in July, when the month's rainfall at Marlborough was 5.60 inches. Harvest weather reasonably good in August and September. Autumn, wet.

Hay was an exceptionally poor crop, very little good hay being made. Wheat yields were well below average. Barley also poor, with much laid corn.

First shipment of grain from the U.S.A. to England. Also first shipment of chilled beef from the U.S.A.

Exploitation of the huge deposits of potash at Stassfurt, in Germany, began.

Barbed wire began to be used about this date.

1876 Wheat 46s 2d per quarter.

Good weather in harvest, but crops suffered from bad weather at sowing-time (in spring and in previous autumn).

Baker records: 'This was a rough wet month, and but little sowing done. Ewes and lambs do very bad owing to the inclement weather, and the hay having been made so bad in 1875, sheep were very poor and weak this spring, many lambs died; in some flocks not more than 50% reared . . . Half the barley not sown by the middle of April. Finished barley sowing, May 9th . . .'

The summer was hot and dry, with the temperature rising to 90°F on August 14th. Heavy rain followed, and September's rainfall at Marlborough was 6.88 inches.

Rainfall in December was 7.20 inches, and the year ended with gales and heavy rain.

Wheat gave low yields, though of excellent grain. Barley poor. Turnips likewise poor, though swedes good.

The potato Magnum Bonum was first put on the market.

1877 Wheat 56s 9d per quarter.

Another wet summer, with poor crop yields.

January rainfall excessive, with much flooding. Spring in general was wet and rather cold, with low temperatures lasting into May. June gave good haymaking weather, with some thunderstorms. July and August, cold and wet. September and October, fairly fine, though with some rain. November, wet again.

Exceptionally heavy storms in the South on July 3rd and 14th. November rainfall at Marlborough was 7.02 inches.

Great gale on October 14th, with great damage to barns, houses and ricks.

Wheat crop well below average. Also barley, with quality indifferent.

The Ayrshire Cattle Herd Book Society was founded.

The Galloway Cattle Society formed.

1878 Wheat 46s 5d per quarter.

A warm, wet summer with crops about average.

Winter was open and rather mild (some autumn corn being sown in January). Several snowstorms at the end of March. April fine and dry, but wet weather early in May lasted till June 17th. July and August were hot, with many thunderstorms. September and early October were fine and sunny, but then the weather turned cool and wet for some weeks.

On June 23rd 3.28 inches of rain fell in 57 minutes in a a tremendous thunderstorm over London.

The wheat crop was above average, but damaged by rain in July and August. Better on the hills than on low ground. A heavy crop of hay was secured.

Poor sowing season in autumn.

The English Jersey Cattle Society formed.

Epidemic of swine fever.

1879 Wheat 43s 10d per quarter.

An unusually wet summer with very bad harvests.

At Selborne, Hampshire, January, February, June, July and August all had monthly rainfalls of over 4 inches, 6.80 being recorded for June and 6.45 for August.

The year started with one of the coldest winters of the century, with much snow. March was fine and dry, but spring was very backward, with cold weather, including snow, continuing well into May. On May 27th heavy thunderstorms began the main period of rain. After three very wet months, September brought rather better conditions, though still with showers. October and November were fine, though with low temperatures. Winter started early, with exceptionally severe frosts in southern Scotland.

A violent gale on December 28th caused the Tay Bridge disaster, when the bridge collapsed and a train was blown into the estuary.

Hay-making was extremely late. Baker records that none was gathered till July 11th, and hay-making ended on August 15th. Harvest did not start till September 4th.

Other comments by Baker include: 'August was very unfavourable. Pastures on clay land are as wet as in the midst of winter. Grass all trodden away, and cattle sink in to their knees . . . Quality of both wheat and barley is wretched in general . . . No corn to sell worth naming, and nobody cares to buy English produce, the quality is so bad. Farmers ruined in every direction.'

The temperatures throughout the summer were exceptionally low.

Sheep and cattle prices declined through scarcity of keep.

The distress caused in the agricultural world was now being aggravated by the vast quantities of grain pouring in from America. These made the rapidly increasing urban population more or less independent of the British harvest.

In this year the first shipment of refrigerated beef from Australia arrived in Britain.

Mechanical separators for milk were introduced.

The British Goat Society formed.

1880 Wheat 44s 4d per quarter.

A fairly good year as regards weather, but corn yields were light.

Much rain in February, but March, April and May were very dry. A backward spring with cold, frosty nights. June and July stormy. August fairly dry, till late in the month; then very wet weather till the end of the year. Snow fell in London on October 19th.

A good barley year, with heavy crops of fair quality, though some was sprouted and stained. Wheat quality good, but yields light. The hay cut was light but not much damaged. An excellent crop of roots.

In this year the potato-growing industry was established in Lincolnshire.

Severe attacks of sheep-rot.

First cargo of chilled meat from Australia arrived at London.

1881 Wheat 45s 4d per quarter.

Another wet summer with a moderate harvest.

This year was distinguished by the worst snowstorm of the nineteenth century, on January 18th and 19th. It came with an easterly gale and, by piling up immense drifts, brought all business to a standstill. Much loss of life, both human and animal. Seventeen men died in Wilts and Berks alone.

Weather remained cold throughout spring, with a frost on May 11th. Rest of May was warm and sunny, but there was a relapse into winter, with a hard frost, on June 9th. July brought very hot weather by day, but a frost on the night of July 28th cut down beans and potatoes. August was mainly wet, but September was a fine, good harvest month. The autumn was cold but mainly frost-free. A great gale on October 13th and 14th did much damage.

The wheat crop was rather below average in yield and suffered much damage at harvest. Barley gave heavy yields of grain that was much damaged. Oats gave reasonably good crops. There was a bumper crop of peas. Turnips filled out well in autumn.

More heavy losses through sheep-rot. In the two-and-a-half years ending with the spring of 1881 sheep-rot is estimated to have killed off 6,000,000 sheep.

1882 Wheat 46s 10d per quarter.

Another very wet year, with deficient harvests.

Winter was mild and dry, and wheat came through well. April brought excessive rains (5.41 inches at Mere, in Wiltshire). Also a violent gale, which did much damage to fruit trees. May mostly dry and cold. with strong east winds. June mostly wet and cold, with frosts on 14th, 16th and 17th. July also wet and cold. August and September stormy, with frequent frosts in September. October excessively wet (6.79 inches on Mere Down). Unusual amounts of rain also in November and December.

Seeds hay, good. Meadow hay, heavy crop but much damaged. Wheat, below average. Barley, rather poor. Oats, about average. Peas, very poor. Turnips, good.

Very little autumn sowing done.

The bacillus of bovine tuberculosis was discovered.

1883 Another unsettled year, with crops below average.

January mild; February very wet; March and April, dry and rather cold. May also rather cool. June hot at first, then thunderstorms. July, stormy, with heavy rains. A very severe hailstorm occurred at Barton, Lincolnshire, on July 3rd. August, also stormy, but with many reasonable harvesting days. Great gale on September 1st and 2nd, followed by dull weather. October mostly fine and cold. November, mostly fine.

Wheat, below average; barley, good; turnips, good.

1884 Wheat 35s 8d per quarter.

A warm summer with good harvests.

A rather stormy and late spring, with frosts and thunderstorms in May. Thunderstorms also in June, July and August, though with intervening fine periods for haymaking and harvest. Early September fine; unsettled later. October, unusually fine and dry. November, damp and cloudy.

Early hay good but light; later hay heavy but much damaged. Wheat excellent. Barley good but discoloured. Turnips good.

1885 Wheat 32s 10d per quarter.

A dry year, with harvests below average.

Winter mild, with no severe frost or snow. Spring cold and backward. May, showery. Drought during June, July and first half of August. Heavy rains in September and October. November, cold and mainly dry. Severe frosts at end of September, and temperatures throughout summer below average.

Wheat was about average but of good quality. Barley rather below average. Oats about average. Peas and spring beans, a failure. Roots, 'the worst crop within living memory'. Hay, average yields but wonderful quality.

Use of basic slag as a manure began.

The Middle White Pig registered as a breed.

1886 Wheat 31s per quarter.

A changeable year with crops about average.

A severe winter. Frosts throughout late January and the entire month of February. Also through much of March. Heavy falls of snow in the north and east; also in Devon and Cornwall. Late and cold spring, lasting, with some warmer spells, into June. Heavy thunderstorms in June. July, August and September—fine spells interspersed with storms, many of them with thunder. October and much of November, stormy, with thunder. The year ended, characteristically, with a blizzard, with deep snow, on December 26th and 27th.

A thunderstorm in South Wales on September 4th was exceptionally severe and caused great damage. Exceptionally heavy rain from May 11th to 13th caused great flood damage in the Severn valley.

Much winter-sown corn killed. Wheat an average crop, but some damaged by blight. Barley excellent, but ripened unevenly. Oats quite good. Hay, average. Peas and vetches, good. Roots, good. Much autumn grass. Feeding stuffs very cheap in autumn.

First measures for the control of anthrax enacted.

1887 Wheat 39s 4d per quarter.

An unusually dry year, with good harvests.

Remarkable for the low rainfall. Only 25.44 inches recorded at Mere, Wiltshire. Only 13 inches in the Fens. The level of Lake Derwentwater fell lower than had ever been previously recorded. Streams in Wales dried up. Many villages, from Lancashire to Sussex, had to be supplied with water from carts.

The early months were quiet and dry. A short period of gales at the end of March. Spring backward and late, with snow on May 20th. Heavy rain in early June; then hot, dry weather, with a few breaks, till the end of August. September, showery. October, mainly fine but cold. November, wet and rough.

Wheat, a heavy crop, Barley, good. Oats, poor. Hay, excellent, especially seeds hay. Roots, poor. Peas, good.

Very little autumn grass, and prices of cattle and sheep consequently fell.

1888 Wheat 31s 11d per quarter.

A rather wet, cool, cloudy year, with comparatively poor harvests.

A rather late spring, though with a good sowing period in early April. May, mainly fine, with a few stormy spells. June, showery, with some thunder. July, stormy and then very wet. (At Mere in Wiltshire rain fell on all but one day in July, totalling 7.77 inches for the month). August gloomy with much rain. September, fairly fine. October, changeable, with rain and sunny spells. November, unusually wet.

On March 24th a remarkable snow-storm, depositing two inches of snow in two minutes, fell at Chepstow, in Gloucestershire. Some of the snow-flakes were nearly four inches in diameter.

On July 15th hailstones up to 2.8 inches in diameter fell at Gloucester.

Both wheat and barley yielded poor crops. Oats reasonably good. Winter beans good, but peas and vetches very bad. Turnips and rape, good; swedes and mangolds, light. Hay, light and in poor condition. Much grass was made into silage. Much autumn grass.

1889 Wheat 29s 9d per quarter.

A showery year with an average harvest.

Frequent snow but few heavy falls in winter. Good sowing weather in March. April showery and spring-like; grass growing well. May, warm and thundery. Early June, likewise. After a wet spell, June and early July brought ideal haymaking weather. Then more showers in July and August, September, mostly fine, with some hot days. October, very wet. November, fine and mild.

A phenomenal year for grass, which gave abundant hay crops. Wheat, barley and most other crops, about average. Livestock prices kept up well.

The Goldthorpe variety of barley originated.

A new Board of Agriculture established.

1890 Wheat 31s 11d per quarter.

Another showery year, rather wetter than 1889. Crops rather below average.

January, stormy. February, mainly fine and mild. March and April, mostly fine, but with some showers. May, a good growing month, but with a damaging frost on the 30th. June, July and August, showery, with much rain. September, an excellent harvest month. October, mainly dry but cold; a heavy snowstorm on the 26th. November, mainly mild but with one or two gales. December, unusually cold; the temperature on the 14th never rose above 21°F, making this the coldest December day on record; snow on several days.

Average crop of hay of moderate quality. Wheat, about average. Barley, rather below average. Oats, excellent. Beans, good. Peas, moderate. Much good grass in autumn.

Pleuro-pneumonia Act came into operation in September and quickly reduced the incidence of that disease.

1891 Wheat 37s per quarter.

A wet summer, but with good crops.

Very severe weather in January, with Fenland rivers frozen and the Thames nearly blocked by ice. February, fine and quiet. March, a very wintry month, with an exceptionally severy blizzard on the 9th. April and early May, cold and backward. Snow fell in parts of England on May 10th and 17th. June, hot and thundery. July, showery but with fine haymaking periods. August, wet. September, fairly fine, but with wet spell in middle of month. October, excessively wet; 9.46 inches of rain at Mere, in Wiltshire; several violent gales. November, fine at first, then more rain. December, wet.

On August 25th/26th 10 inches of rain fell in the Lake District.

Wheat, good; barley and peas, good. Oats, light. Some corn not harvested till first week of November in Wiltshire. Roots, rather small. Cattle and sheep prices very low in autumn.

A Herd Book formed for the South Devon breed of cattle.

1892 Wheat 30s 3d per quarter.

A rather dry year, with poor harvests. Baker calls it 'a most disastrous year'.

Some snow and frost in January and February. The Highlands of Scotland had an exceptionally heavy snowfall from January 5th to 9th, followed by a severe flood when the thaw came. March, wintry. A very late spring. Very little rain. Late May, June and July were thundery, with fine spells. August started with good weather, but heavy rain fell later in month. September was a good harvest month. October, showery. November, mild and misty.

Nearly every month till August was deficient in rainfall, with less than an inch of rain in March and May.

Grass short, and hay crop extremely light. Wheat, below average. Barley, good, with excellent quality. Oats, very poor. Peas, light. Beans, variable. Roots, variable.

Store stock prices were very low in autumn.

Epidemic of foot-and-mouth disease.

Passing of the Contagious Diseases of Animals Act, giving power to slaughter diseased animals, with compensation.

1893 Wheat 32s per quarter.

A year of drought. Crops rather below average.

A mild winter, followed by an early spring. Much rain in February, but one of the finest Marches on record. April also fine, calm and sunny; temperatures rising to 70°F. No rain till May 14th. During this spring drought many places in

Kent and Sussex had no rain for 50 days. Showers in late May, and some thunder in June. July, showery. August, mainly fine. September, showers and fine spells alternating. October, first showery, then fine. November, mostly mild and damp, but with one week of gales and heavy rain.

On August 10th 20.09 inches of rain fell in 35 minutes in a tremendous thunderstorm in Lancashire.

Wheat, average crop of good quality. Barley and oats, poor and unevenly ripened. Turnips and swedes, average. Mangolds, patchy. Prices of sheep and cattle fell in autumn, but were kept higher than they would have been because the drought was less severe in the north.

1894 Wheat 22s 10d per quarter.

A showery year, with harvest rather above average.

A fairly mild winter. Showery but cool in spring. Many cold storms and night frosts in May. Warm with thunderstorms in June. Showery and dull for most of July, and also much of August. September mostly fine, but with some storms. October rather cold. November, very wet (rainfall 7.16 inches at Mere, in Wiltshire).

Wheat, average, with much straw; barley, heavy crop of poor quality; oats, good. Hay crop heavy and of good quality. Roots variable. Autumn keep plentiful, and prices of store stock fairly good.

1895 Wheat 23s 1d per quarter.

A rather hot, dry year, with a deficient harvest.

January and February, frosty, with some snow; March, cold but mostly dry. The Thames was blocked by ice-floes for about a week in mid-February. On February 11th a temperature of −17°F was recorded at Braemar, Aberdeenshire. A severe gale, particularly in the Midlands, on March 24th. April, mostly cold, with heavy storms. May, rather hot, with drought. June, drought. July, drought till 19th, when violent thunderstorms broke. August and September, hot and sunny, with storms. A temperature of 87°F recorded in London on September 24th; and for 18 days in September the temperature exceeded 70°. October, alternately fine and stormy; latter part of month very cold, with snow. November and December, mild and damp, with gales.

Wheat, poor (much destroyed by cold winter): barley,

some average, but late-sown crops failed to ripen; oats, light. Hay crop, good. Also much good straw. Turnips, average.

Herd Book formed for Lincoln Red cattle.

First milking machines introduced.

1896 Wheat 23s 1d per quarter.

A dry year but a wet harvest. Crops good.

January and February mild and dry. March rather wet. April dry and rather warm. May, almost absolute drought; vegetation badly affected. June, dry and very hot, with a few thunderstorms. July, excessively hot, with a few heavy thunderstorms; otherwise dry. August, one fine week for harvest; otherwise wet. September, unusually wet; rainfall 7.70 inches. at Mere in Wiltshire. October, wet at beginning, then fine. November, fine and becoming cold. December, wet at first then cold, with some snow.

Wheat, an excellent crop. Barley and oats, below average. Turnips, variable; failures where no storms occurred. Hay crop, very light. Shortage of keep in autumn, and cattle and sheep consequently cheap.

Cockle Park Experimental Station established in Northumberland.

The bacillus of contagious abortion of cattle was isolated.

The value of basic slag as a fertilizer for grassland was demonstrated.

1897 Wheat 26s 2d per quarter.

Again rather dry until harvest, then stormy. Crops about average.

A mild winter and wet spring. Vegetation very forward. May mainly dry, with a hard frost on the 12th. June and July, dry, though the rainfall figures were boosted by several heavy thunderstorms in June. August, showery. Also September. October and November, fine with October unusually warm. December, wet.

Wheat just below average, but quality good. Barley, poor. Oats average. Early-sown roots were good; late ones a failure. A heavy hay crop, mostly well made.

Heavy mortality among sheep in early spring, in spite of the mild winter. Probably due to a mineral deficiency.

1898 Wheat 30s 2d per quarter.

A hot, dry summer, with good crops.

A January spring. Low rainfall for next three months, particularly March. Heavy snowfall on February 21st. April, showery and chilly. May, unusually wet and cool, with over 5 inches of rain. June, cool, with showers and dry spells alternating. July, very dry. August and September, also dry; September particularly hot. October, fine till the 13th, then almost continuous rain. November and December, changeable and stormy.

Wheat, rather above average, though some was damaged by rain in May. Barley, average. Oats, a splendid crop. A very heavy hay crop, well made, but not much aftermath. Peas, good. Roots, below average. Price of sheep fell off in autumn, with shortage of keep.

1899 Wheat 34s per quarter.

Another hot, dry summer. Crops average.

A mild winter, followed by a backward spring. May, mainly dry. June, hot and sunny, with several thunderstorms. July, very dry and hot, with a few violent thunderstorms. August, the same. September also dry, but becoming cooler. October, mainly fine with heavy rain at end of month. November, heavy rain, then fine and quiet. December, a period of frost in middle of month; mostly damp and dull.

One of the earliest harvests on record. Nearly all corn harvested by middle of harvest. No delays by bad weather. Autumn unusually fine and quiet.

Wheat, average but of excellent quality. Barley, good. Oats, rather light. Hay crop, below average but in excellent condition. Very little aftermath.

The Large Black Pig Society formed.

The Burger oil-driven tractor appeared.

1900 Wheat 25s 8d per quarter.

A showery year, with rather light crops.

A January spring. February snowy at first, then incessant rain. March, cold and dry. Also April. May and June, mainly dry but some showers. July, very hot and mostly dry, with a few showers. August, showers and fine spells. September, fine, dry and warm till last week, then showery. October, dry in middle of month, showers early and late. November, mild and showery. December, mild and wet.

Wheat, average, good quality but very short straw. Barley, below average. Oats, variable, some very poor. Potatoes, heavy crop. Roots, good. Hay rather above average; autumn grass plentifiul. A bumper year for fruit.

Pleuro-pneumonia of cattle eradicated about this date.

Exceptional rainstorms on December 30th, with 3½ inches recorded at Worcester.

The English Dexter Herd Book founded.

1901 Wheat 26s 10d per quarter.

A dry summer, with average crops.

A frosty but not particularly severe winter. Then a backward spring. Drought set in about the middle of April. May, June and July, fine spells, growing increasingly hot, with intervals of showers and thunder. August mainly fine, but with two short periods of thunder showers. September, cooler but still showery. October, light rain. November, mostly calm and dry, with one gale. December, much rain.

The North had much more violent weather than the South. A blizzard raged throughout Scotland, northern England and Wales on March 29th. Violent gales in same areas from November 11th to 13th.

The highest barometric reading ever recorded was 1055 mm at Aberdeen.

Wheat, average, with short straw. Barley, variable but quality not good. Oats, poor. Hay, rather light but well made. Winter beans, average; spring beans, poor. Turnips, late crops fairly good; early ones damaged by turnip fly. Potatoes, good. Mangold, good.

Harvest began early and was mostly finished during August in the South.

The long-lost papers of Gregor Mendel, the Moravian monk who discovered the principles of heredity, were found and published.

1902 Wheat 27s 6d per quarter.

A mild winter and showery summer. Harvest a little above average.

After a mild winter, spring was rather late. Cold till the middle of May, with night frosts. June to September, inclusive, showery, with some short bright spells. October, mostly quiet, with dry and stormy spells alternating. Very little extreme weather all the year.

Wheat, good. Barley, good, but stained. Oats, heavy. Roots, good. Hay, good, though some damaged by rain. Grass, abundant. Potatoes, much disease.

1903 Wheat 28s 2d per quarter.

A wet summer, with crops about average.

The year began quietly, but violent gales occurred in February and March. A gale which swept the country, particularly the North and Scotland, on February 26th and 27th was exceptionally severe. April, cold and frosty. May, showery at beginning, then dry and hot. June, variable, with very wet spell in middle of month. July, a good haymaking spell in middle of month; otherwise stormy. August, showery. September, mostly fine, but a tremendous gale on the 10th. October, excessively wet; 8.57 inches of rain at Salisbury. November, fine and quiet. December, showery.

Wheat, late and damaged, but about average. Barley and oats, much the same. Hay, heavy cuts, and much of it of good quality. Much grass. Roots, average. Fruit, almost a complete failure.

The first milk-recording societies were formed—these in south-west Scotland.

1904 Wheat 26s 9d per quarter.

A showery summer, but with reasonable fine spells. Harvest, about average.

Another year without extremes of weather. Rather cold but without excessive rain for first three months. April, mainly fair. June, mainly fine, but some thunder. July, fairly hot, but with some thunder. Also August. September, first half wet; then dry. October, showery till the 11th, then fine. November, dull and showery. Great snow storm in North in middle of month. December, damp, foggy and rather cold.

Wheat, rather below average. Barley and oats, about average. Hay, rather better than average. Plenty of grass. Roots, average.

1905 Wheat 28s 4d per quarter.

A fine harvest, with yields rather above average.

A fairly open winter, with much sowing in February and

early March. Heavy gale, March 15th and 16th. April, mild at beginning, then colder. May, cold and dry June, mainly dry, but with two or three heavy storms. July, mainly dry and hot, but with a few storms. August, showery to the 11th, then fine. September, fine but cooler. October, dry and rather cold. November, storms and frosts alternating; severe gale on 26th. December, fine and dry.

Wheat, excellent. Barley and oats, average. Turnips, average. Hay, a fairly good cut, though later crops affected by dry weather.

A separate Herd Book formed for Dairy Shorthorns.

Dr E.S. Beavan, at Warminster, produced the parents of Plumage-Archer barley.

1906 Wheat 29s 8d per quarter.

A dry but rather cool summer. Good crops.

First three months, rather wet, especially January. Latter half of March and all April, dry and cold. A backward spring. June, mainly fine, with some thunderstorms. July, dry and cool. August, a few showery days but mostly warm and dry. September, very hot for a few days at beginning of month; otherwise fine and pleasant, with some gentle rains. From August 31st to September 3rd the temperature reached over 90°F on four successive days. On September 1st it rose to 96°F. October, wet. November, showery, December, mild at first, then frosty. Heavy snowfall in Scotland, December 26th to 30th.

Wheat and barley, excellent crops. Oats, rather below average. Hay, average cut. Roots, quite good, but rather stunted by drought.

1907 Wheat 28s 3d per quarter.

A rather dry summer with excellent crops.

First three months of the year, dry, with rather low temperatures. April, showery but rather cold. May also cool, with rough storms at beginning. June, cool and dull, with a little snow in the South on the 12th. July, fine but not hot; showers at beginning of month. A series of severe thunderstorms in the south on July 21st and 22nd. August, dull and cloudy, with a little rain early in month. September, warm bright weather nearly all month. Very little rain. October, excessive rain; 8.13 inches at Salisbury. November and December, mild and damp.

Wheat, a rather heavy crop of low quality. Barley and oats, very heavy. Hay, a heavy crop, but much of second quality. Roots, unusually good.

Passing of the Smallholdings Act.

Founding of the National Cattle Breeders' Association.

1908 Wheat 30s 7d per quarter.

A showery summer, with rather poor crops.

Cold in early months of year. Temperature 18°F in the south on January 5th. Heavy snowstorm on March 3rd. April, cold for most of month. Very heavy snowfall on April 25th, nearly a foot deep on level fields. Melted by the 28th. May, showery and warm. June, hot, with some thunder. July, fine at first, then showery; warm. August, fine for first three weeks; then stormy. September, mainly showery. October, mild and quiet. November, the same, except for a gale on the 22nd. December, mild and damp, then growing colder.

Wheat, rather below average. Barley, rather poor. Hay, average. Roots—mangolds, good; swedes and turnips, below average.

The Lincolnshire Farmers' Union, which was to develop within a year or so into the National Farmers' Union, founded by Colin Campbell.

Dr H. Hunter produced the celebrated Spratt-Archer barley.

1909 Wheat 32s per quarter.

A stormy summer, with good crops of rather low quality.

Cold in January and February. February very dry. March, wet. April and May, variable, but on the whole a backward spring. June, showery, with some heavy thunderstorms. July, stormy and rather cool. August, first fortnight fine and bright; then wet. September, a fine harvesting spell in middle of month: otherwise rainy. October, wet. November, mainly fine and cold. December, showery, with severe gales on 2nd and 3rd, especially in north.

Wheat, oats and barley, heavy crops but much damaged by rain. Hay, a light crop, badly made. Roots, fair.

The British Friesian Cattle Society formed.

1910 Wheat 36s 11d per quarter.

A cloudy rather wet summer, but with very good crops.

Frost and snow in January; much rain in February. March, cold and mainly dry. April, showery with cold winds. May, showery; cool to begin, then warm and summerlike. June, stormy; heavy thunderstorms on the 5th and 9th July, showery and dull. August, mostly dull and fine, with showers at beginning and end; very little sun. September, fine and sunny, but rather cool; only .24 inches of rain at Salisbury. October, quiet and rainy. November, all sorts of weather. December, wet and mild.

Wheat, average. Barley and oats, very good. Roots, good. Hay, good crop, but later cuts much damaged by rain.

The King Edward potato introduced.

At this date 90% of the land of England and Wales was cultivated by tenant farmers.

1911 Wheat 31s per quarter.

A year of drought, with crops above average.

January, dull and damp. February, cold and dry. March, cold, with showery spells. April, mainly dry, with cold northerly winds, but showery at end of month. May, mainly fair, with a showery spell in middle of month. June, first half hot and dry; then showery. July, almost absolute drought, with great heat. August, hot and dry. September, drought continued till the 13th, when the drought broke. During the drought the temperature reached 100°F at Greenwich on August 9th. Autumn was showery, and December excessively wet.

Wheat and barley crops were well above average; oats about average. Hay was a good crop, but rather light. The pastures were burnt up. Mangolds gave an average crop, but heat, mildew and insects stunted other roots and made them almost a total failure.

Baker considered this 'the hottest and driest summer in memory of man'.

The Majestic potato introduced.

1912 Wheat 34s 9d per quarter.

Many gales in January. Dry in late January and early February; renewed gales in March. April, very dry. Summer wetter than usual, and with less sunshine. A few hot spells, maximum temperature in London being 91°F from July 12–15. August wet, with excessive rain in the eastern counties on August 26, over 4 inches falling in 24 hours. Floods in East Anglia, London and Scotland. Some dry periods in September and October, but skies remaining cloudy. A very severe gale in Scotland on December 24th, causing many casualties.

Harvest, rather below average, particularly of potatoes.

The first sugar beet factory in England was established at Cantley in Norfolk.

1913 Wheat 31s 8d per quarter.

A dull year, with an open winter, wet spring, and a very dry summer which was, however, neither sunny nor warm. Winter mild.

January and March were months of boisterous gales, and April and May were more windy than usual. Numerous thunderstorms in May, and a severe storm on June 9–10. July, August and September were quiet months October and November were stormy, with exceptionally severe gales and tornadoes in western districts on October 27th. Some loss of life caused in Glamorgan.

Rainfall was slightly below normal, except in south-west England, where there was some excess. Temperature slightly above normal (highest 85°F in London on June 16th and 17th). Summer deficient in sunshine. Autumn unusually mild.

Nearly all crops about average. Much good late hay made, and a good aftermath of autumn grass.

Potato root eelworm first appeared in Britain.

The beginning of official milk recording in Britain.

1914 Wheat 34s 11d per quarter.

A mild, sunny year.

A mild beginning to the year, with few frosts in January and February and much rain in February. Floods in Thames valley. Drought in April, May and June. At King's Somborne, Hampshire, only 0.46 inches of rain between April 11th and June 30th. Some storms elsewhere in latter half of June, and July very stormy. Highest temperature—94°F in London on July 1st. August, a warm, dry, sunny month. Storms in early September, but drought from September 18th to October

25th over much of England. Storms in late October, November and early December, with some snow in the North in November.

Harvest above average for wheat and barley. About average for oats and potatoes, and also for root crops, which were somewhat affected by droughts. Early hay cut light but well made.

1915 Wheat 52s 10d per quarter.

A cool, dull year, with alternating spells of drought and rain.

January and February brought heavy rain and some snow. Floods in mid-February. March, April, stormy. May, rather quiet, with an early stormy period. A drought which began in the south in mid-May lasted till the third week of June. June was, on the whole, hot, with temperature rising to 90°F at Cromer on the 8th; though sharp frosts occurred at night between the 18th and 21st, doing much damage. Weather broke in latter part of June, and July was unusually wet and stormy. Wet spell lasted till about mid-August, but second half of month, and first half of September, dry. Severe frosts in November—it being reckoned one of the coldest Novembers on record. December, very stormy, with severe gales from 22nd to 27th.

Winter wheat, rather better than average, but spring cereals affected by drought in May and June. Hay light, but early crops well made.

1916 Wheat 58s 5d per quarter.

A dull, wet year.

Violent gales in January, with tremendous downpour in Scotland, causing landslip between Fort Augustus and Fort William, which disrupted rail traffic for days. March was the wettest since 1841. A violent storm on the 27th and 28th, with heavy rain and sleet in some places, and snow in others, did much damage, uprooting trees, flattening buildings, etc. Over 100 greenhouses wrecked in Guernsey. Much flooding, especially in central England.

Dry spells in late April and in second half of May. Also in mid-June and from 10th July to 10 August. Shorter dry periods in September. June was unusually cold (the coldest June in Dublin for 50 years); growth of vegetation much retarded and damaged by frost. This cold spell lasted until about July

15th. Heavy rains from October onwards. Sunshine deficient for most of year. 8 inches of rain in Scottish Highlands on October 8th.

Winter wheat, below average. Spring cereals about average, though some rather below. Hay about average, though somewhat damaged by storms.

1917 Wheat 75s 9d per quarter.

A cold year, with a wet summer.

Very cold in January and February. Many English rivers, from the Nene to the Severn, frozen for the first time since 1895; and rapid streams such as the Wye also frozen. At Sheepstor, on Dartmoor, 91 consecutive days of frost recorded—the longest spell of frost since 1855. March more unsettled, but temperature still low. April, north-east winds, with very heavy snowstorms in Scotland and the North. Cold till middle of month. May, warm and sunny. June, warm but very stormy. 250 mm (9.56 inches) of rain which fell at Bruton (Somerset) on the 28th and 29th is a record for the British Isles. July fine but cool, until towards the end, when rain fell. August, excessively wet. September, average. October, very stormy, with severe gales on 24th-26th; exceptionally wet in Scotland. November, dry and rather warm; December, dry but colder.

Crops average throughout, though barley slightly below. Much damage to quality through harvest rain. Some good hay made. Roots good.

Corn Production Act passed, guaranteeing minimum price for wheat and oats and a minimum wage for farm labourers. Agricultural Wages Board set up.

In Northumberland one farmer lost two-thirds of his ewe flock before lambing; and many lambs were deliberately killed at birth because their mothers were too weak to nurse them.

1918 Wheat 72s 10d per quarter.

A rather wet year.

January, cold, with much snow. Severe frosts in Scotland, with river Dee and parts of Loch Lomond frozen. Temperature on January 8th –3°F at Peebles. Sharp rise in temperature towards end of month. Mildest February since 1903. March colder but mostly dry (though with blizzards in the north). April, dull, cold and wet. May opened cold but became warmer towards middle of month, ushering in a period of thunderstorms, with much rain and hail. June opened and closed warm, but middle of month had sunny days and cold nights. July, sunshine and heavy showers, with normal temperature. August, changeable, but warm and dry on whole, with fair periods of sunshine. September, very rainy and cold. October, dull, damp and unsettled, though with little rain or sun. November, damp and foggy, with severe frosts at times. December, mild and unsettled, with much rain.

Rainfall in general above average, except in Scotland. The September rainfall in Lancashire and the West Riding of Yorkshire was the greatest ever recorded. 41½ hours of continuous rain began at Meltham (Yorks) on September 14th.

Winter wheat, above average yield. Spring cereals, rather below average. Hay cuts good, but damaged by rain. Potatoes and roots, average. Fruit crop an almost complete failure everywhere, owing to severe frosts on April 3rd and 23rd.

1919 Wheat 72s 11d per quarter.

A generally dry year.

January, dull and wet, with cold weather at beginning and end of month. February, first half of month very cold, with east winds; second half, mild and rainy. March, exceptionally cold, with much snow—the coldest March in Scotland for over 60 years. Precipitation (either as rain or snow) above average everwhere. April, changeable and unsettled, with frequent showers of rain or hail, and with a very heavy snowstorm on April 27th and 28th. May, sunny, warm and abnormally dry. June, hot and dry till 19th; then cool and rainy. July, warm but with unsettled spells. August, a sunny month, with many hot days, but with moderate rainfall. September, fine, dry and quiet. October, dry, sunny but with many frosts. November, exceptionally cold, with severe frosts and much snow and hail. December, dull, rainy and mild.

Temperatures for the whole year below average, particularly in Scotland.

All crops below average, probably owing to cold late spring and the drought in May and early June. Wheat particularly affected. Hay crop deficient.

The Board of Agriculture became the Ministry of Agriculture.

Professor Sir R.G. Stapledon appointed first director of the Welsh Plant Breeding Station at Aberystwyth.

The Essex Saddleback and Wessex Saddleback pigs registered as breeds.

1920 Wheat 80s 10d per quarter.

A dull year, with a cool, damp harvest.

January, mild, wet and windy, except for first ten days, which were cold. February, mostly mild and dry, but with snow on 19th and 20th and several foggy days. March, unusually mild. April, unsettled, showery and rather cold. May. cool and rainy till the 20th, then warm and thundery. June, frequent thunderstorms and no very hot days. Finer in Scotland than in England. In the Midlands and South exceptional frosts between the 4th and 10th did much damage. July, dull, wet and cool; unusually cold in Scotland. August, dull and cool, with widespread ground frosts. September, weather improving, but with much dew and fog. October, mostly sunny and bright. November, fine but foggy. December, mild, except for wintry spell in second week.

Spring crops, about average. Winter wheat, rather below average. Roots, good. Hay, good in quantity but not in quality. Plenty of autumn grass.

Agriculture Act passed, regulating cereal prices.

1921 Wheat 71s 6d per quarter.

An abnormally dry year, with almost unparalleled drought.

January, very mild but rainy. Severe gale in south on 18th. February, exceptionally dry—one of the driest Februaries on record. Relatively warm. March, mostly showery and mild. April, mostly sunny and dry, with cold spell in middle of month. May, fine and dry. Severe frost on the 5th June, absolute drought in most places, especially in the south. The driest June since records were kept. July, drought continued unbroken in most places. August, cooler and rather unsettled, though with rainfall below average. September, mostly fine and warm, but with thunderstorms on the night of 11th and 12th. October, hot and dry for most of month. November, still dry but cooler. December, mild, with rain everywhere except the south-east.

The weather in north-west Scotland provided a marked contrast to that of most of the rest of the country, with rather more than average rain. Taking England and Wales alone, less than half the normal amount of rain fell for the six months February to July. Only 9.29 inches of rain for whole year recorded at Margate, the lowest total ever for any part of the British Isles.

Winter wheat was exceptionally good. Oats about average. Barley rather below average. Potatoes and roots, much affected by drought. Hay was scarce though well-made. Autumn grass very short. Most harvest was finished exceptionally early.

Repeal of the Corn Production Act.

1922 Wheat 47s 10d per quarter (effect of the repeal of the Corn Production Act in 1921).

A cool, unsettled summer.

January, changeable, with much rain and snow. February, mild and unsettled. March, began in same way but switched to cold, dry weather. April, mostly cold and wet. May, mostly dry, though with thunderstorms at beginning and end of month; a hot spell began about the 20th. June, warm and sunny till middle of month; then cool and unsettled. July, cool and unsettled, with much rain and little sun. August, the same. September, two fine spells, otherwise rainy and dull; cool. October, exceptionally dry, though with low temperatures; snow widespread on 28th and 29th. November, quiet, cloudy and dry, with much fog. December, mild and sunny to start; many gales later.

Spring was unusually cold and backward. At Bolton (Lancs) it was the coldest April since 1887. Over much of England the rainfall was more than 20% above average, due chiefly to great cyclonic rains in early August. By contrast, it was the driest October within living memory.

Winter wheat, slightly above average. Barley and oats, below average. Potatoes and roots, above average—quite good crops. An abundance of grass but hay much damaged by rain.

Serious outbreak of foot-and-mouth disease.

Summer Time Act passed.

Milk and Dairies Act (passed 1915) came into force, designating different grades of milk.

1923 Wheat 42s 2d per quarter.

Another dull, wet year.

January, mild and sunny, apart from a few cold days at beginning. February, mild and very wet. March, first week mild and unsettled; then a cold fortnight; then unusually warm. April, warm at first; then cool, May, after a few warm days the month was cool and wintry, with many storms. June, cool and dull but dry. July, warm and thundery at first; then cool, wet and windy. August, a fine spell in middle of month in south-east; otherwise gales with much rain. September, warm and sunny at first; then unsettled period with storms and gales. October, boisterous winds and heavy rain. November, very changeable, with much rain but also much sunshine. Gales. December, foggy at first; variable later, with much snow in places.

The wettest year since 1916. Rainfall excessive almost everywhere. On the island of Islay rain fell on every day from August 12th to November 8th—89 days. The number of rainy days in Connemara exceeded 300, which is nearly a world record. Borrowdale in Cumberland had 238 inches.

Winter wheat gave a yield above average. Spring cereals were about average, as also were potatoes. Roots and grass, good. Hay, a good cut, though later crops damaged by rain.

Another serious outbreak of foot-and-mouth disease.

Devon Closewool Sheep registered as a breed.

1924 Wheat 49s 3d per quarter.

Another very wet year.

January, warm and unsettled; much heavy rain. February, unusually dry, with low temperatures and cold east winds. (Severe blizzard on January 8th–9th.) March, mostly sunny with cold nights, but cloudy and mild towards end of month. April, cool, cloudy and wet, with gales and thunderstorms. May, very wet and thundery. June, dull and unsettled, though with a warmer, drier spell at end of month. July, mostly heavy rains and thunderstorms, but with fine spell in middle of month. August, cool and wet, though with a few fine spells. September, much heavy rain, though with a few fine spells. October, unsettled and rainy, though drier at end. November, mostly warm and dry. December, warm and unsettled.

One of the wettest years on record. Seven consecutive

months, April to October, all had above average rainfall.

Rainfall in general ranged from 130% to 180% of average. For Hampstead, London, the rainfall for July was 7.60 inches, a record only twice surpassed for London. 9.40 inches fell on August 19th near Cannington, Somerset.

Crops were about average for yield, though wheat was rather above average. Much damage was, naturally, caused by the weather.

Another serious epidemic of foot-and-mouth disease.

1925 Wheat 52s 2d per quarter.

A normal year, the outstanding features being a very dry June and a cold December.

January, mild and stormy. February, mild, wet and windy. March, persistent dry north winds, April, showery and cool, with some bright periods. May, cool and cloudy, with much rain at times; a warm, sunny spell in middle of month. June, warm and very dry. July, the dry spell of June continued for first three weeks; then cooler and showery. August, cloudy and unsettled, with short dry spell in middle of month. September, cold and windy with much rain; but fine spell in last week. October, mild, but with one cold spell. November, mild for first week; then dry and cold. December, wintry, with some snow, until last week, when milder. The cold spell, in fact, lasted from November 8th to December 26th, and was quite severe.

Spring crops about average, but wheat excellent, on the whole. Some very good hay was made. Root crops average, though the early ones affected by the June drought. Much grass about in autumn.

1926 Wheat 53s 3d per quarter.

A mild, dull year.

January, wet and mild, with cold spell in middle of month. February, unsettled, dull and mild, with cold spell in middle of month. March, mostly dry and mild, with cold spell in middle of month. April, mild and dull, with much rain in places. May, cool and changeable. June, cloudy till middle of month; then warm and dry. July, mainly fair and warm. August, unsettled, though with fine periods. September, warm and mainly dry. October, unsettled and mild till middle of month; then very cold. November, very wet, with frequent gales. December, dry and sunny.

This year, in contrast to the previous series, spring crops did better than winter ones. Barley and oats were above average. May frosts, however, affected roots and particularly potatoes, which gave a yield below average. Much good hay made.

Fowl pest, or Newcastle disease, first occurred in Britain.

1927 Wheat 49s 3d per quarter.

Another very wet year.

January, mild and wet, except for a wintry spell in third week. February, foggy and cold till 19th; then mild and wet. March, wet and mild. April, unsettled till 18th; then brief warm spell; then very cold. May, a quiet month; warm and sunny. June, cold, unsettled and with much heavy rain, except for warm, fair period in middle of month. July, thundery, unsettled and dull. August, unsettled and wet, except for brief fair periods at beginning and end of month. September, cool, very wet and windy, except for brief fine period at beginning. October, after first few wet days, fine and dry, until last week, which brought gales. November, changeable, with mild and wintry spells alternating. December, mostly dull and wintry; much snow 24th to 27th,—a notable blizzard.

The summer was not only wet but very deficient in sunshine. The period June to September was the wettest since 1879. July was distinguished by very severe thunderstorms. The frosts of April and early May were unusually severe, and in Scotland it was the coldest June for 60 years.

In spite of the apparently inclement weather, most crops yielded above average. Little good hay was made. Sugar beet, recently established on a large scale, gave a disappointing harvest.

1928 Wheat 42s 10d per quarter.

On the whole, a wet year, but with dry, warm periods in central and eastern England in summer.

January, mild, stormy, with about twice normal rainfall over much of Britain. February, mild and wet, with gales and floods during first half; then dry and frosty. March, mild and wet, with one cold spell in mid-month. April, changeable and showery, with considerable fair periods. May, warm and sunny at beginning and end; cold, cloudy spell in middle. June, mostly unsettled and cool, with much rain. July, warm and sunny, with cool wet spells at beginning and end. August, mainly fair sunny and dry in south and east; much rain elsewhere. September, fair and sunny. October, mild and dry, but with cold spells at beginning and end. November, quiet and rather cold at first; unusually wet and stormy after the 10th. December, variable.

A stormy year, apart from the unusual dry summer in the south and east. In spite of the variation in conditions from place to place, the harvest was in general well above average. This applied to nearly all crops, except sugar beet. Not a good season for early haymaking, but meadow hay produced useful crops.

Formation of the Agricultural Mortgage Corporation.

1929 Wheat 42s 2d per quarter.

A sunny year, with temperature and rainfall about normal.

January, quiet, cold and dry. February, dull, cold and dry; coldest February since 1895 in many places; temperature −1°F at Ross-on-Wye. March, sunny, quiet and dry, with cold nights. April, dry with cold north-east winds. May, mostly cool and unsettled; second half, finer, but many periods of heavy rain. June, dry and sunny but rather cool. July, mostly dry and sunny. August, mainly fair and dry in England and Wales; unsettled and wet in Scotland. September, mainly fine and warm. October, unsettled, wet and windy, but with bright periods. November, mild, much rain. December, mild and stormy, but with frequent sunny periods. Severe gales.

Several places in the south and east had the driest August for many years. In south Wilts only 10 inches of rain fell in first nine months of the year. Many also recorded exceptional hours of sunshine. November was excessively wet.

An excellent harvest, with nearly all crops yielding above average. The average barley yield, of 17.8 cwt per acre for England and Wales, was the highest of the century until yields in general began to climb in the late 1930s and 1940s. Plenty of hay.

The last Sussex team of oxen (at Birling Manor, East Dean, Sussex,) was dispersed.

Agricultural land de-rated.

1930 Wheat 34s 3d per quarter.

A wet year, with persistent rain.

January, mild and wet, with some severe gales. A particularly violent gale on the 12th. February, mostly cold, quiet and dry. March, wintry with much frost and snow from 9th to 23rd; otherwise mild and wet. April, mild and wet at first; then wintry and dull; last week sunny and dry. May, generally dull and wet in south and east; dry elsewhere. June, sunny, warm, with frequent thunderstorms. July, unsettled, dull and wet. August, mainly cool and wet, with a few fine spells. September, persistent rain. October, November and December, mild and mostly wet; severe gales in early November.

Weather caused much damage to crops. Wheat and barley much below normal in quantity and quality. Oats, about average. Roots did very well. Hay poor, except for earliest cuts. Much grass in autumn.

1931 Wheat 24s 8d per quarter.

Another wet, dull year. No really hot days in summer, and many cold spells.

January, first half, cold and foggy; second half, mild, wet and windy. February, much rain and snow, with gales. March, cold and quiet. April, dull and unusually wet; one warm spell in middle of month. May, mostly wet, cloudy and mild. June, first three weeks dull and wet; then fine. July, dull and wet, with frequent thunderstorms. August, wet and cool in south and east; sunny and dry in north and west. September, mostly cold and dry, but with some warmer spells. October, dry and sunny till 21st; then severe frosts. November, very mild, wet and cloudy. December, mild, dry and cloudy.

Year noteworthy for deficiency of sunshine. December unusually warm. Violent thunderstorm in West Country on May 27th did exceptional damage in north Devon. 9.6 inches of rain in two days in early November near Brecon.

A rather poor harvest, with much loss through bad harvesting weather. Wheat about average for quantity, but barley, potatoes and beet below average. A bad hay-making season.

Agricultural Marketing Act passed.

Formation of the Land Settlement Association.

1932 Wheat 25s 4d per quarter.

Another dull year, with little sunshine though without excessive rain.

January, very mild; wet in north and west, dry in east. February, quiet, dry and sunny. March, dull, cool and mostly dry. April, wet, cool and dull. May, cool, cloudy, and much rain; wettest May for many years in England. June, dry and sunny. July, dull and wet, with widespread thunderstorms. August, warm and dry on whole, but many local thunderstorms. September, mainly wet and cloudy. October, very unsettled and rather cold. November, dry and dull. December mild and mostly dry in England; wet in Scotland.

Noted for deficiency of sunshine. Month by month records for dullness were broken. Otherwise an undistinguished year.

Crops also undistinguished, all being about average, though much good hay was made.

The Import Duties Act was passed, imposing duties on agricultural imports, except from the Dominions and Colonies.

Wheat Act passed, imposing quota of home-grown wheat on millers.

Rights of Way Act passed.

1933 Wheat 22s 10d per quarter.

An exceptionally dry year.

January, mild at first; then cold. February, wet and mild at first; then colder, with heavy snow 23rd to 26th, great drifts. March, warm and sunny; April, drought; 23 consecutive days without rain; cold spell 17th to 22nd, otherwise warm. May, warm and dull, with many thunderstorms. June, warm and sunny. July, warm and sunny, except for the north, where many thunderstorms. August, warm and very sunny. September, warm and sunny, except in south-east England, where much rain fell. October, warm and dull. November, dry and cool. December, dry and cold.

Sunshine records were the best of the century (except for 1911). Temperatures well above normal (maximum 94°F, Cambridge, July 27th). The period June to September was the warmest since before 1881. December was exceptionally cold (the coldest since 1890 in much of England and Wales).

An exceptionally good crop of wheat. Oats above average;

also sugar beet. Barley about average. Hay cut light, and little autumn grass.

Establishment of Milk Marketing Board.

1934 Wheat 20s 9d per quarter—the lowest average reached during the Depression.

A mild year, with both drought and floods.

January, wet with sunny intervals; many gales. February, very dry; mild in north, cold in south. March, unsettled, rain variable but above average, rather cold. April, mostly wet; heavy snow in Scotland, 5th to 10th. May, mostly wet; temperatures normal, with warm spell in middle of month. June, dry and warm, but with heavy local thunderstorms. July, the same. August, unsettled, with rainfall and temperatures about average. September, warm and sunny, but with heavy rain in places. October, dull and mild, but with two cold snaps; very cold and heavy snow on 30th and 31st. November, dry, quiet, mild, much fog. December, very mild and wet.

June, July and September were exceptionally warm. December was the warmest for over 60 years. Rain fell mostly in thunderstorms.

An exceptionally good crop of wheat again. Sugar beet also good, and potatoes above average. All root crops flourished. Spring-sown cereals about average. Good hay cut and much autumn grass.

Milk first supplied free to school-children.

Farmers Weekly first published.

1935 Wheat 22s 2d per quarter.

A variable year, mild on the whole and with rainfall above average. A wet and stormy autumn.

January, rather mild and dull. February, very mild and wet, with gales. March, mostly dry and mild, but a cold spell with heavy snow 8th to 11th. April, mainly wet, mild and dull. May, mainly very dry and sunny, with very cold spell 12th to 19th. June, first three weeks very wet; then warm, sunny spells. July, dry and sunny, but not hot. August, warm and dry, but spells of rain in southern England. September, very wet, with severe gales. October, dull and wet. November, dull and wet but mild. December, foggy, then very cold spell 17th to 24th.

A notable feature of the spring was a heavy snowfall, with

severe cold, between May 13th and 17th. The period June 20th to August 22nd was unusually warm, sunny and very dry. September broke records for rain, the rainfall being 300% of average in parts of England and Wales. A very severe hailstorm did much damage in Northamptonshire on September 22nd. September and October gales were unusually violent and widespread.

Most crops about average, though wheat again well above average. Sugar beet also good. Much damage was done to fruit and vegetables by the May frosts. The late hay cut was good.

First Attested Herd scheme introduced.

A.G. Street's book, *Farmer's Glory* published.

British Sugar (Subsidy) Act passed.

Restriction of Ribbon Development Act passed.

1936 Wheat 30s 9d per quarter.

A variable year, with extremes of rain and drought.

January, very wet; cold except in south-east England. February, cold and wet, with east winds. March, mild but very cloudy. April, cold, with persistent north winds. May, very dry. June, severe thunderstorms with heavy rain; cool spell at beginning; warm later. July, cool, cloudy and wet, with heavy thunderstorms locally. August, mostly dry; cool at first but becoming very warm towards end of month. September, dull, with frequent storms. October, mostly dry and sunny, but with strong winds in second half of month. November, dull and wet in most places; strong winds to 18th, then foggy. December, rather mild and sunny, but with some gales.

Sunshine deficient throughout year. Extensive floods in January. Long drought in April and May. Unusually frequent thunderstorms in June and July. Good harvest weather in August. Very windy autumn.

Most spring crops about average. Wheat considerably below levels of past few years. Sugar beet, good. Hay cut, rather deficient.

Formation of the British Sugar Corporation.

Passing of the Tithe Commutation Act.

1937 Wheat 40s per quarter.

A wet year in England and Wales, though dry in Scotland.

January, mild and very wet; frequent gales. February,

very wet and rather mild, but with notable snowstorm on 27th and 28th. March, very cold, with frequent sleet and snow; much rain in England, with severe flooding in the Fens. April, very dull, with much rain. May, dull and wet in much of England; dry and sunny in Scotland and Wales; warm spell in last week. June, dry, though with thunderstorms; fairly warm. July, dull, much rain, chiefly in thunderstorms. August, warm, and mostly sunny and dry; absolute drought in many places; but thunderstorms elsewhere. September, variable, with fair amounts of rain; cool spell in middle of month. October, dry and dull. November, dry foggy. December, cold, with much snow, frost and fog.

An unusually sunless year. A very cold March; and unusually heavy precipitation during first five months. June and November, on the other hand, were abnormally dry, in some places, though much depended on local thunderstorms.

Crops on the whole below average. Barley particularly poor. A good hay cut. Roots, average.

Subsidy payments for barley and oats introduced. Also grants for lime and basic slag, and T.T. milk legislation.

1938 Wheat 28s 11d per quarter.

An extremely variable year.

January, mostly wet and mild, with many gales. February, dry; mild at first; then cold. March, dry and very mild. April, very dry. May, frosts for first ten days; mostly rainy and cool. June, very windy, but mostly dry, though very wet in Scotland. July, cool, dull and stormy. August, remarkable series of thunderstorms in first twelve days; warm, then cooler. September, dull and rather warm, with some rain. October, unsettled and rather mild, frequent gales in west and north. November, mild, wet and windy. December, cold spell from 18th to 26th, with severe frost and snow.

The spring drought was exceptional, April probably being the driest since 1727. In the August thunderstorms, from the 1st to the 12th, Torquay had 6.39 inches of rain, mostly in 9 hours, and Lexden in Essex had 2.25 inches in 60 minutes. Much large hail fell.

An exceptionally heavy wheat crop. Barley also much above average, and oats a little above average. Potatoes quite good, but sugar beet light. Hay, light.

We are indebted to THE TIMES for permission to reprint the following article by their Meteorological Correspondent on the weather of 1938, which is headlined 'An Astonishing Year':

'For extent and variety of atmospheric extravaganzas in Great Britain the year 1938 may safely be said to have had no peer within memory. Every month has been marked by eccentric behaviour of one kind or another.

'On the night of January 25th-26th, the finest auroral display of the century was seen, not only from all parts of this country but even as far south as Madeira.

'Gales swept over our islands on as many as twenty-one days between January 8th and 31st; several of the storms attained full hurricane force, and on the 15th, 23rd and 29th gusts of 100 or 101 m.p.h. were registered locally in the west and north.

'February was notable for further frequent tempests and for heavy snowfalls, with drifts up to 8 feet deep in Kent from the 13th to the 17th.

'March outstripped all records for sustained warmth nearly everywhere, the mean temperature in some places exceeding the previous highest for that month by 4 or 5 degrees. Astonishing variations of rainfall occurred in Scotland; Loan (Inverness-shire) received 50.03 inches, one of the greatest monthly totals ever logged in Britain, whereas some places in East Lothian had less than ¼ inch.

'Meanwhile, districts south of the border were in the midst of a remarkable drought; over England and Wales as a whole, February, March and April appear to have been drier than any other three consecutive months since 1785. In London (Kew) their total yield of rain was 0.7 inches (15% of the average). Various localities, especially in the south-western counties, came through April without a measurable shower.

'Winter returned in May, when some of the keenest frosts known so late in the season wrought immense havoc amongst the prematurely advanced fruit crops. At Rickmansworth and Thetford the air temperature fell to 18 degrees on May 8th.

'June entered with the most violent summer gale that can be remembered on the south coast, the wind speed reaching 88 m.p.h. at Calshot on the 2nd. Scotland had twice the normal rainfall during this month, though over much of south-east England there was a deficit of 60% or more.

'July distinguished itself by bringing snow to parts of the Scottish Lowlands as well as to Cheshire, between the 5th and the 19th, and a destructive tornado to the Chiltern hills on the 7th. A humid heat wave with shade temperatures approaching the 90 mark in the south around the beginning of August heralded Devon's worst thunderstorm deluge for at least seventy years.

'On the 4th Torquay had 6.4 inches of rain—nearly all of it in about nine hours. Some of the accompanying hailstones were the size of small walnuts. On August 21st an exceptionally early frost visited Hertfordshire. A warm but dull and foggy September sent London's thermometers to between 80 and 83 degrees as late as the 23rd.

'October passed without a dry day locally in the north, and was the wettest for over seventy-five years at Lancaster. November, like March, broke all records for continued warmth over much of the country; the 5th, with temperatures up to 70 or 71 degrees here and there in the south, was by far the most summerlike day known after the end of October. A gust of 180 m.p.h. occurred near Pembroke during a hurricane on the 23rd.

'So genial did the weather remain until mid-December that birds were nesting and trees budding, but on the 18th a frost unequalled since 1929 was suddenly imported from Russia and within 48 hours skating was possible on shallow waters. Then came a snowstorm which was commonly the heaviest since February, 1933, and locally since Christmas, 1927.

'This recapitulation of a strange year's outstanding natural phenomena would not be complete without a reference to the minor earthquake which shook London and the Home Counties on June 11th.'

1939 Wheat 21s 5d per quarter.

A wet year in the south and east, but drier in west and north.

January, wet, with some severe gales. February, mild. sunny in most of England; wet in Scotland and Wales. March, dull, with alternating dry and wet spells, and one cold spell late in month. April, sunny and warm. May, dry, becoming warm. June, sunny and warm at first, becoming cooler and less settled. July, dull, cool, many thunderstorms. August, mostly dry, but some rain and thunderstorms; cool at first, becoming warm. September, dry and mostly warm. October, rather cold; much rain in south-east; sunny in north-west. November, very mild and dull; rain and gales. December, dry and rather sunny; cold in second half of month.

All crops better than average. Sugar beet gave highest yields yet in Britain. Wheat good, though not quite up to 1938 standard. Roots good.

At this date, the outbreak of war, Britain was importing over 70% of its food.

County War Agricultural Executive Committees set up.

1940 Wheat 42s 10d per quarter.

A very cold winter was followed by warm, dry weather early in summer and unsettled, rainy weather in late summer and autumn.

January, intense frost; sunny early, then much snow. February, dull and very cold to start; wet in south later. March, cold at first but temperature rising; dry at first, then unsettled. April, wet and rather warm. May, dry, sunny and warm. June, very sunny, warm and dry. July, unsettled and cool; frequent thunderstorms. August, very dry. September, first week warm, then cooler; mostly dry. October, dull, unsettled, mild; many gales. November and December, mostly wet and windy; mild at first, then colder.

Outstanding features were the warm, dry weather in May and June (the warmest June in Scotland for over 80 years); the dry August; and the very wet November.

Wheat and barley, average crops. Oats rather above average. Potatoes, sugar beet and most roots, above average. Much good hay made.

In February a freezing rain caused severe ice formation on trees, fences and grass and did much damage to young forest plantations.

Local Defence Volunteers formed; some agricultural workers carried their rifles to work.

1941 Wheat 62s 10d per quarter.

A dull, cool year, on the whole. Mostly dry in the west. Unusually cold from January to May.

January, very cold, with frequent snow. February, cold with frequent snow; heaviest fall for over 100 years in north England and south Scotland. March, rather cold, with north-east winds and much snow and rain. April, dry, dull and mainly cold, with north-east winds. May, cold, especially in the south, with many severe frosts. June, dry and cool, but becoming warmer later. July, warm, becoming hot, with many thunderstorms. August, cool and unsettled, with much rain. September, dry, dull and very mild. October, mild at first, becoming cold. November, dull and rather mild; dry in west; wet in east. December, dry and mild, with much fog, but a severe gale on the 6th and 7th.

Wheat about average. Spring cereals rather deficient. Roots and potatoes, quite good. A useful hay crop. Much damage to fruit by May frosts.

Combine drills began to be used in Britain.

1942 Wheat 68s 2d per quarter.

A year of extremes. Unusually cold with much snow in first quarter; periods of drought and heavy rainfall later.

January, severe frost and much snow. Dull. February, very cold and dry; mostly cloudy; heavy snow in first week. March, cloudy and cold; several snowfalls. April, sunny and rather warm; last three weeks dry, with strong east winds. May, sunny, except locally in south; some rain. June, dry. with alternate warm and cold spells. July, unsettled and cool, with showers. August, mostly cloudy and cool, with thunderstorms; a warm spell at end of month. September, unsettled; very heavy rain in England, with much flooding. October, mainly dull, unsettled and mild. November, quiet, cool and dry, with much fog. December, mild and unsettled, with much rain; some severe gales.

Exceptionally heavy snow in first three months. Much fog and cloud. Rather cool.

A year of excellent crops. Wheat, barley and oats all much above average, though quality damaged at harvest, which lasted a long time. Potatoes and sugar beet above average, though not much so. Early hay good; later crop damaged. Much autumn grass.

1943 Wheat 69s 7d per quarter.

A mild, wet year on the whole, though relatively dry in the east.

January, excessive rain; cold to 10th, then mild; severe gales at end of month. February, very mild, with S.W. winds; mostly dry. March, mild and dry. April, warm and mostly dry; some gales. May, alternating wet and dry spells; rather warm. June, mainly unsettled, with frequent local thunderstorms; last week drier. July, dry and very sunny, becoming very warm in last week. August, dull and wet in north and west; dry in south and east. September, unsettled and mostly wet, with many thunderstorms. October, dull, mild and mostly wet, with frequent fog. November, mainly unsettled, but rainfall not high; some snow and sleet. December, mostly dry and cold, with frequent fog.

A good year for most crops. Wheat and barley yielded above average; oats, potatoes and sugar beet about average. The averages, however, are now beginning to create difficulties, as they improve annually with the rapid increase in use of chemical fertilizers and more efficient farming methods. Before 1940, for instance, an average (for the whole country) of 19 or 20 cwt per acre of wheat was exceptional; by 1950 it was reckoned as low.

1944 Wheat 63s 11d per quarter.

A dull year with an unsettled summer.

January, mild and cloudy, with much rain and numerous gales. February, began mild, but cold later; dry, but heavy snowstorm 26th and 27th. March, very dry, apart from heavy snowstorms in Scotland. April, warm, with first three weeks unsettled. May, dry and sunny in England, dull and wet in Scotland; heavy thunderstorms late in month. June, mainly unsettled, dull and cool; much wind. July, dull and rather warm, with showers. August, mainly warm and sunny till 19th, when heavy thundery rains. September, cool and rainy, with two short dry spells. October, rather cold with frequent rain. November, very wet, with strong winds. December, mainly cold, with severe spell around Christmas; snow in middle of month.

Nearly all crops about average, with wheat slightly above. Poor hay harvest. Much autumn grass.

1945 Wheat 64s 10d per quarter.

After an intensely cold start, a mild and rather dull year.

January, frost and much snow; very heavy falls in Scotland; severe gale on 18th and 19th. February, very mild and wet. March, very mild and mostly sunny and dry. April, mostly sunny and dry, but a cold spell late in month. May, unsettled and wet, with many thunderstorms and very wide range of temperatures. June, unsettled and wet, with average temperatures. July, changeable and warm; severe thunderstorms 13th to 15th. August, mainly sunny, warm and dry. September, dull and rather wet. October, dry and unusually warm till 20th, when unsettled weather began. November, dry, mild, cloudy and quiet. December, unsettled and rather mild, with stormy period 16th to 29th.

March, April and November were exceptionally dry months. October was the warmest since 1921. Sunshine deficient.

All crops above average, barley considerably so. Much damage was done, however, by the thunderstorms of mid-July, making the grain of indifferent quality and harvesting difficult. Hay also much damaged.

1946 Wheat 66s 9d per quarter.

A very wet year.

January, rather cold with much rain in west; two dry frosty periods at beginning of month. February, mild and wet for most part, with heavy flooding following rain on 7th and 8th; cold spell from 21st. March, dry; cold to 17th, then warm. April, warm, sunny and dry, particularly early in month. May, dry and sunny in north; wet and cool in south. June, dull and wet, with frequent thunderstorms; rather cool. July, warm at first, then cool; several series of severe thunderstorms. August, cool, unsettled and wet, with some gales. September, mostly wet with many gales. October, very dry, quiet and dull. November, mild, dull and very wet. December, unsettled except during cold dry spell, 15th to 21st.

An average year for most crops. Sugar beet yielded well above normal—an average of 10.6 tons per acre, which was the highest recorded in England to date. Wheat slightly above average. A difficult harvest, with much damage to corn. Poor haymaking.

Hill Farming Act passed.

1947 Wheat 75s 4d per quarter.

A year of extremes; with periods of intense cold and excessive heat; floods, drought and heavy snowfall.

January, unsettled, with mild spell in middle of month; severe wintry spell for 23rd onwards, with heavy snow. February, very intense and prolonged cold, with heavy snowfalls. March, extremely cold with heavy snow in first half; then milder but still with much precipitation as rain or snow; severe floods in consequence. April, very wet, with frequent gales. May, unusually warm; mostly wet in England, but dry in Scotland. June, very hot at first, then warm: unsettled with frequent thunderstorms. July, rather warm, with cool spell 5th to 11th; thunderstorms. August, very hot, sunny and with little wind. September, very warm; dry and sunny in south and east, wet and dull in north and west. October, dry, quiet and mild. November, unsettled and rapid changes of temperature. December, mainly dull and dry.

February was the coldest month at Oxford since 1815; and the coldest month for all England and Wales since 1895. August was the warmest since before 1881. March was one of the wettest ever recorded.

Deficient harvests all round. Crops yields in general back to the levels of the 1920s.

In the blizzards from January 23rd to middle of March, about 6 million sheep and some 30,000 store cattle were lost.

Agriculture Act, basis of farm policy for many years to come, was passed.

Harry Ferguson, in conjunction with the Standard Motor Company, began making tractors equipped with hydraulic three-point linkage.

YEAR OF EXTREMES

Reproduced by kind permission of The Times

'It is a commonplace that Nature does not repeat herself. Thus it is not surprising that the rainfall of 1947 differed from that of any earlier year, both in the distribution over the country and in the incidence throughout the year.

'In spite of an infinite variety, Nature achieves a balance, so that the rainfall of these islands is confined within fairly definite limits. The year 1947 was outstanding, however, in that the limits of monthly rainfall were extended beyond those experienced within living memory, giving the wettest March on record in England and Wales, the wettest April in Scotland, the driest August over the country generally, and the driest October over England and Wales. In spite of these extremes, the annual totals at most stations were about normal, no station having less than 73 per cent. or more than 121 per cent. of the average.

'The pattern of the distribution of the annual amounts in inches is similar each year, since the mountainous districts of the west always receive more rain than the lowlands of the east. Thus at Princetown the totals in 1946 and 1947 were 106·87 inches and 79·48 inches respectively, while at Margate the corresponding totals were 24·92 inches and 18·99 inches.'

THREE DRY MONTHS

'Over large areas stretching across the centre of the British Isles the rainfall exceeded the average. In the north-west, in parts of Ross-shire, annual totals failed to reach 80 per cent. and it was there that February, March and August were unusually dry. The largest areas with less than 80 per cent. were in the south-east, due in this case to the dry weather during April to December. At Lowestoft the rainfall failed to reach the average in each of these nine consecutive months.

'During the summer half-year the rainfall was appreciably less than average over most of England and Wales, but in excess of the average over Scotland. This followed a winter half-year with the opposite distribution—namely, wet in England and Wales but drier than usual in Scotland. As a result England and Wales suffered more than Scotland both from curtailments of water supplies at the end of the summer and from widespread flooding in March.

'Annual totals for some representative stations are shown on p. 198.

MONTHS REVIEWED

'*January* brought the average rainfall to most districts, apart from the north-west of Scotland, which had less than 75 per cent.

'*February*, with the preponderance of winds from the east,

gave more than the average rainfall in the east, where from Spurn Head to Berwick-on-Tweed amounts generally exceeded 150 per cent., while in parts of the west of Scotland, e.g. at Glen Quoich, Glencoe and Ardgour, there was no measurable rain at all, an event unprecedented in this mountainous region.

'*March* continued with easterly winds in Scotland, and the same type of rainfall distribution prevailed there as in February, but over England and Wales winds from the south and west predominated and the month was windier than usual. Parts of the Western Highlands of Scotland received only one-quarter of the average, while nearly the whole of England and Wales received more than twice the average, with more than three time the average from London to Hereford, around Shaftesbury and to the south of the Wash. Widespread flooding occurred which was unprecedented in many districts. It was due not only to the prolonged rains but to the melting of accumulated snows and the small percolation into the frozen ground.

'*April* brought three times the average amounts over much of the English Lake District, southern and central Scotland.

'*May* was generally wetter than usual, but gave less than the average in the eastern half of England and north-western half of Scotland.

'*June* and *July* gave normal amounts in most districts, but there was an abrupt change at the end of July.

'*August* gave no measurable rain at all in many parts of Scotland, in the Fen District and Kent. Most stations recorded less than 10 per cent. and only a few, in south-west England, received more than one-third of the average.

'*September* brought general rains, the amounts varying from half the average in East Anglia to twice the average in Sutherlandshire.

'*October* was generally dry, with less than one-tenth of the average in the east of England and amounts reached half the average only in west and central Scotland. The total rainfall from June to October over England and Wales was only 8·8 inches, even less than the previous low record of 9·1 inches for the similar period of 1921.

'*November* gave more than the average in the west of England and Wales and over most of Scotland, but there was less than half the usual amount around London.

'*December* was generally dry, but there was more than the average from London to the east coast between Dover and Lowestoft, while the English Lake District and Western Highlands of Scotland received less than half the usual rain for that month.'

DR. JOHN GLASSPOOLE

1948 Wheat 94s 6d per quarter.

An exceptionally sunny and warm spring followed by a cool, rainy, dull summer.

January, relatively mild, though with widespread snow in north. February, much snow. March and April, unusually warm and sunny. May, also sunny, but with some unsettled periods. June, July and August, dull, cool and wet. September, October and November, very mild, and unsettled, apart from cold spell in middle of October. December, very mild at first, then colder with much snow.

January was the wettest on record for over 80 years for England and Wales. March had the highest March temperature for over 100 years, at Greenwich. April was the sunniest on record. At Edinburgh, August, with 9.4 inches of rain, was the wettest month since September, 1785.

Crops well above average. Wheat, the highest average recorded to date—20.7 cwt per acre. Barley and oats also, at 19.4 cwt and 17.7 cwt per acre respectively. Sugar beet excellent. Difficult harvest again spoiled quality of much grain. Haymaking likewise difficult.

Selective weedkillers for cereals were introduced.

1949 Wheat 104s 7d per quarter.

A dry, sunny summer.

January, February and March, dry and mild on the whole, with very little snow. January the driest in England since 1911. February and March had only half the normal rainfall over much of England. April and May, wet months, but not sufficient rain to restore the average. June, dry. July, very dry apart from local thunderstorms, some of which were violent. August, mostly dry. September, mostly drought, apart from heavy thunderstorms, particularly in west. October, a wet month. November, rainy. December, dry in south and east; wet in north and west.

The sunniest year at Kew since 1880, when records started.

Driest year since 1933 in England and Wales. October, however, was the wettest since 1903.

A year of bumper grain harvests. The wheat average climbed to 22.4 cwt per acre; the barley to 20.6 cwt per acre —both records. Sugar beet and potatoes, however, fell below average, and all root crops were badly affected by drought. Some good hay was made, but grass became scarce, and milk yields fell off in summer and autumn.

1950 Wheat 119s 7d per quarter.

A wet year.

January, a very dry month. February, frequent heavy rains, also snow, with much flooding. March, generally mild and rainy. April, much rain. May, mainly dry. June, mainly dry. July, unsettled with much rain and many thunderstorms. August, unsettled, with frequent thunderstorms; cool. September, wet and stormy. October dry, with period of frost at end of month. November, wet and mild. December, dry and cold.

January was the driest in England since 1907. February was the wettest since 1869 (with the exception of 1923). Rainfall from July to September was excessive.

Crops less than in 1949 but still well above former averages. Much damage caused by heavy rain at harvest. Root crops exceptionally good, though potato blight was prevalent. Sugar beet achieved unprecedented yield of 12.3 tons per acre average. Much good hay made.

Area eradication scheme introduced for bovine tuberculosis. Diseases of Animals Act collected in one omnibus act all previous legislation on subject.

1951 Wheat 126s 4d per quarter.

A fine, favourable year for farm crops, though rain caused damage during harvest. January, February and March were unusually mild, with exceptional rain in February. June and July were dry months, July bringing several very warm spells. August was stormy, but several fine periods in September aided the harvest. October was unusually dry, calm and mild. November, wet.

All crops yielded above average; wheat, barley and oats all about 2 cwt per acre above the average for the previous ten years. 7.9 tons per acre for potatoes was a record to date.

All root crops also gave bumper crops.

Barley acreage increased by 7% over 1950.

1952 Wheat 129s 4d per quarter.

Another favourable year, with a relatively fine summer.

January, rather mild. February, cold and dry. March, mild and showery. April, mild and dry. May, warm and dry; absolute drought from 13th to 29th. June and July, warm and dry; 27 days without rain in July. August, changeable with some heavy rain. September, heavy rain in last ten days. October to December, changeable.

Once again, all crops yielded above average and all were above the levels of the previous year, with the exception of barley, which was slightly below. Root crops were also lower than in 1951, and potatoes about the same.

In spite of slight variations in yield, the total barley harvest was now increasing annually, owing to the steady increase in acreage. This year the total barley harvested amounted to 2,065,000 tons, as against 1,747.000 tons in 1951.

Exceptionally severe gales in Scotland in January. The year of the Lynmouth flood disaster, when on August 15th floods caused by 10 inches of rain on Exmoor destroyed the centre of Lynmouth. A smoke fog in London on December 6th caused many deaths and much trouble among livestock at Smithfield Show.

Serious outbreak of foot-and-mouth disease in Scotland, and of anthrax among pigs in England.

1953 Wheat 136s 10d per quarter.

Yet another favourable year, though a stormy harvest caused delays and waste.

January, dry and rather cold. February, very dry, apart from second week. March, exceptionally dry. April, drought and storms alternating May and June dry, with some showers. July, stormy. August to December, mostly dry with sunshine above average, but several wet spells notably in mid-September and second half of October. Temperatures above average.

Yields of wheat, barley and oats per acre surpassed all records, as also did those for potatoes, sugar beet, swedes and mangolds. The actual figures per acre were as follows: wheat, 24 cwt; barley, 22.6 cwt; oats, 21.2 cwt; potatoes,

8.7 tons; sugar beet, 12.8 tons; swedes, 17.8 tons; mangolds, 25.7 tons.

This was the year of the North Sea floods, which on January 31st caused catastrophic flooding along the east coast, from Lincolnshire to Kent, and even worse flooding in Holland. Prime cause was a severe north-north-west gale combined with high tides.

It was also the year when myxomatosis first appeared among rabbits in England.

1954 Wheat 114s 4d per quarter.

A very wet year. After a cold winter and a dry, sunny though rather cold spring, the period May to November brought excessive rain and was also cool. Frequent thunderstorms occurred. Haymaking was a wash-out, and cereal crops were slow to mature. Rain interfered with harvest as well as haymaking, and also with root harvesting in autumn. One spell of fine weather in the last week of August was interposed between two long periods of rain. There was a heavy snowfall on Scottish mountains on September 17th, and incessant rainfall for three days in Scotland and northern England from October 17th to 19th. Exceptionally heavy rains with violent gales in November.

The yields of all crops were lower than in the excellent harvest of 1953 but were nevertheless all above the average for the previous ten years. Wheat gave 22.6 cwt per acre; barley 21.4 cwt; potatoes, 7.9 tons.

Myxomatosis spread rapidly.

Atrophic rhinitis first recorded in Britain, having been imported from Sweden in Landrace pigs.

1955 Wheat 99s 4d per quarter.

A fine, warm summer, with light rainfall, apart from a few thunderstorms. July and August, remarkably sunny and dry (although local thunderstorms provided exceptions, as at Upwey in Dorset on July 19th, when 9.5 inches fell in 24 hours, one of the heaviest falls ever known). After the first week of September a rainy spell set in.

Once again, all records for yields were broken for grain. Wheat gave 26.7 cwt per acre; barley, 25.4 cwt; oats, 22,5 cwt. Potatoes and roots, however, were badly affected by drought and were below average. A wonderful haymaking season. Also an exceptional year for harvesting broad red and other species of clover.

1956 Wheat 108s per quarter.

A cold late spring was followed by warm weather in early summer; then exceptionally wet spell till late September, when finer weather enabled some of the cereal crops to be salvaged. February was exceptionally cold, surpassed in this century only by 1947. Rainfall from February to May was very deficient, the smallest for nearly 100 years. July produced new rainfall records for many places.

Grain yields were down on those of the very favourable year of 1955 but above the ten-year average. Wheat gave 24.7 cwt per acre; barley, 23.8 cwt. The hay harvest was most disappointing, with low yields and poor quality. Root yields, on the other hand, were high; sugar beet gave 12.3 tons per acre, and potatoes 8.3 tons.

In this year the Ministry of Food was combined with the Ministry of Agriculture and Fisheries.

1957 Wheat 92s 7d per quarter.

A late, backward spring, characterised by cold, dry weather. This merged with a June drought and heat wave. It was the driest April for many years, and some places had an unbroken drought of 40 days in April and May. The temperature in London on June 29th rose to 96°F, the highest in June for over 100 years. This dry spell was broken in July, and thereafter the summer was cloudy and wet, with frequent thunderstorms and much heavy rain. Dry and relatively mild weather returned in October and lasted to December.

Wheat gave a good yield (25.3 cwt per acre), but otherwise all crop yields were down considerably on those of the previous year, though mostly they were above the ten-year average. They were affected by the cold spring, by the excessively hot June and by the rains of harvest. Haymaking was good, most of it being finished before the rains came.

Potato production fell by 23%, partly due to reduced yields and partly to reduced acreage.

1958 Wheat 92s 3d per quarter.

Another wet summer. Heavy and persistent rain, with many

severe thunderstorms, characterised the summer months. Temperature and rainfall average for January and February, but rainfall was below average for March and April. May and June were very stormy, as were also August, September and October, the September rainfall being nearly three times the normal. November was a dry month, but December wet.

All crop yields were below those of the most recent favourable year, 1955, but still above the ten-year average. Wheat yielded 24.5 cwt per acre; barley 22.8 cwt; oats, 19.8 cwt; sugar beet 13.3 tons; potatoes, 6.9 tons. The sugar beet yields are the exception to the general run, being the highest recorded to date. Most roots were good, but the hay cut was poor and delayed, with much damage by rain.

Silage-making was increasing in Britain, and in this year forage-harvesters began to come into general use.

A hailstorm at Horsham, Sussex, deposited the largest hailstones on record for Britain, one weighing 6½ ounces.

1959 Wheat 90s per quarter.

A fine, warm summer. A mild spell in February saw the temperature rise to 66°F in London on the 28th. Spring was dry, and a drought which began on May 4th lasted throughout most of the summer in many districts, though some areas had thunderstorms and infrequent outbreaks of rain. A temperature of 84°F was recorded in London on October 3rd, and the summer drought did not break there until October 10th.

The warmest year of the century in Scotland.

A record year for all cereal crops, in spite of some being adversely affected by the drought. Wheat gave 28.8 cwt per acre; barley, 22.8 cwt; oats, 22.cwt; all these being all-time records to date. Although root crops were retarded by lack of rain, both potatoes and sugar beet gave above-average yields. The hay crop was in excellent condition though rather light.

The larger acreage of barley, together with higher yields, resulted in an increase in total production of 778,000 tons (27%).

1960 Wheat 90s 4d per quarter.

A rainy summer, on the whole. July was unusually cool

and wet, with several night frosts recorded. A spell of exceptionally rainy weather which began on July 4th lasted, in most districts, till September 6th. Rainfall during this period was excessive, with violent storms causing much damage to cereal crops and delaying ripening. Much grain was, however, salvaged in September.

In spite of the rain, crops yielded well above the ten-year average, though they were lower for grain than in 1959. Wheat gave 28.4 cwt per acre; barley, 25.0 cwt; oats, 21.4 cwt. The sugar beet yield was the heaviest ever recorded—16.8 tons per acre. Potatoes and roots were about average, though there was much waste with potatoes through difficulties at lifting-time. The hay harvest, except the very latest cuts, was good and reasonably heavy.

A tornado swept parts of Wiltshire and Gloucestershire on August 27th. Highest August rainfall for England and Wales since 1939. Severe floods around Exeter in late September, where 5 inches of rain fell in four days. October was one of the dullest and wettest ever recorded. November was warmer and sunnier than usual, though also rainy.

1961 Wheat 89s 8d per quarter.

Changeable. The spring and early summer were cool though mainly dry. As heat increased in July and August, so storms became more frequent. Severe gales in autumn. January cool and dull; temperature 10°F at Grantown-on-Spey on the 16th. February and March, mild and dry. April, wet but mild. May, drought throughout middle of month; severe frosts at end. June, dry and sunny. July, mostly dull and cool, but temperature reached 93°F in London on the 1st August to October, mild and changeable. November, cool and dry. December, cold.

Wheat yielded rather less than in 1960, but barley yields (26.1 cwt) and oats (23.1 cwt) were both up. The barley harvest had now risen to 4,430,000 tons, topping the 4 million ton mark for the first time. Both potatoes and sugar beet yielded well above the ten-year average. The hay crop was generally good.

In this year negotiations began for Britain's entry into the European Common Market.

Temperature fell to 9°F at Edinburgh on December 27th.

Heavy snowfall in southern England on December 31st. Temperature then fell to –15°F at Grantown-on-Spey.

1962 Wheat 88s 1d per quarter.

A very cold spell early in year, followed by rather cool, late spring, with much wind. Many night frosts in June, a remarkably low temperature of 20°F being recorded in Norfolk on June 3rd. July, August and September were changeable, with periods of strong winds and heavy rains alternating with warm, dry weather. Much rain in autumn. Very severe spell with blizzards in last few days of year.

A bumper harvest for most crops. Wheat gave the unprecedented yield of 34.8 cwt per acre; barley, 29.0 cwt; oats, 25.2 cwt. Early potatoes were much damaged by late frosts, yet the total yield of 9.2 tons per acre was a record. Sugar beet considerably lower than in previous two years, and the hay cut, though gathered in good condition, was rather light, owing to the cold dry spring.

Free Calf Vaccination Scheme against Brucellosis introduced.

1963 Wheat 88s 1d per quarter.

Exceptionally severe winter. Snow which fell in late December, 1962, was still on the ground in mid-March, though augmented by several subsequent blizzards. Very low temperatures, and cold east winds. Thaw began in southern England on March 4th, and March was a wet month.

After a mild spring and warm, sunny weather in early June, changeable and cool weather set in. The rest of the summer was mostly cool and wet, with warm, sunny spells at the end of July and in mid-September. Autumn was rainy, with frequent gales. November was exceptionally wet and mild.

Crop yields were below those of 1962 but not much so, and still well above the ten-year average. Wheat yielded 31.3 cwt per acre; barley, 28.3 cwt; oats, 24.6 cwt. Potatoes and roots were also down a little.

Temperatures fell to 5°F at Stansted (Essex) on January 23rd and to –8°F at Braemar on January 18th. 22 inches of level snow at Princetown (Dartmoor) on February 6th. The sea froze in January off the Kent coast, and ice-floes occurred on the Thames in London. January was the coldest month of the century.

The severe winter caused considerable losses of ewes in some parts, with effects on the lamb crop.

1964 Wheat 93s 9d per quarter.

On the whole, a fine, favourable summer. April was cloudy and wet, and spring sowing late. May, warm and showery. June, unsettled, with much rain. July, mostly dry and warm, though unsettled in north. August, warm and dry in south, though with a few unsettled periods. September, warm, dry and sunny, except for one wet spell. October, dry and cold, with some rain. November, dry and mild. December, cold and very wet.

An excellent harvest on the whole. Wheat, 33.7 cwt per acre; barley, 29.4 cwt; oats, 26.4 cwt. Sugar beet gave the high yield of 14.2 tons; and potatoes were above average with 9.2 tons. A heavy crop of hay in reasonably good condition, but pastures became short of grass late in summer.

1965 Wheat 97s 2d per quarter.

A wet, dismal year, with excessive rain and deficient in sunshine. After dry, cold weather in February and March, spring sowing was early, and April was reasonably warm and sunny. The weather deteriorated as the year progressed, and persistent rains characterised the summer and autumn. Much damage was done to crops, and weeds flourished more than for many years.

Barley gave a very heavy yield, the heaviest to date (29.9 cwt per acre), though the quality was in general poor. Oats also gave heavy crops, but wheat was below the previous year's level, though still good. Potatoes gave the heaviest yield ever—10.6 tons per acre; and sugar beet and roots were also heavy. The hay crop was almost a complete failure, and harvesting was very frustrating and expensive.

1966 The winter proved very unfavourable for work on the land, cultivations being almost at a standstill throughout January and February. This was due to the wet conditions, the soil seldom having a chance to dry out. In March a fine spell enabled sowing to get well forward, but any work which was not completed then had to wait until the end of April, as the early weeks of April were dull and wet. In particular, the middle of April saw wide-spread snow and unusually low temperatures with the consequence that there were heavy losses of lambs and ewes in hill flocks, and the spring bite of grass was very late. May brought excellent growing weather, and some good early cuts of hay were taken in the first week or so of June, but after that came a period of unsettled thundery weather, which delayed the hay harvest and spoiled much of it. In July more hay was ruined by heavy thunderstorms, and the weather in general was cool and unsettled. Some storm damage was reported in cereal crops. Similar conditions prevailed over most of the country during August, when the grain was slow to ripen and harvest proceeded at a snail's pace. September, however, proved to be a splendid harvesting month, with warm and sunny weather most of the time. The quality and yield of all cereals was excellent, though no records were broken. Some farmers were able to take a good second cut of hay, and grass was plentiful. October, November and December were, on the whole, dull and wet, with October and November colder than usual but December mainly mild. The harvesting of potatoes, sugar beet and other roots proved difficult, though the yields were good. Autumn cultivation fell into arrears, and less autumn corn was sown than usual. Stubble fields were dirty with weeds, particularly with black grass. Supplies of winter keep appeared to be adequate, and cattle and sheep started their winter in good condition.

WEATHER PATTERN OF ONE OF THE WETTER YEARS OF THE CENTURY

Reproduced by kind permission of The Times

'Rainfall was greater than average in 1966 over each country of the United Kingdom. Years as wet as 1966 occur every five or six years on average over England and Wales. The rainfall excess over Scotland was not remarkable, but Northern Ireland experienced the third wettest year of the present century (1954 and 1929 having been wetter).

'After a dry January over the major divisions of the country ('dry' in this context is a relative term: a month which appears dry relative to average may be wetter in absolute terms than other months with greater than average rainfall), a prolonged wet period was experienced from late winter to mid-summer.

'Although March was dry over England and Wales, and April dry over Scotland, the period February to June was the wettest such period since 1951 over England and Wales and the wettest since comparable records began in 1869 over Scotland and in 1900 over Northern Ireland.

'The second half of the summer was much more favourable over Scotland and Northern Ireland, and in fact July to September combined were the driest such months since 1959 over these countries. Over England and Wales, however, the summer continued generally unpleasant, as much for sustained coolness as for rainfall, until mid-September. Not since the summer half-years (April-September) of 1931 and 1932 have there been two consecutive summer seasons as wet as those of 1965 and 1966 over England and Wales. It is perhaps not encouraging to note, with 1967 holidaymakers in mind, that three consecutive very wet summers occurred in 1930, 1931 and 1932.

'Among the largest and smallest annual rainfall totals recorded were 173 inches at Styehead, Cumberland, and 20 inches at Newmarket, Suffolk.

'The year began mild and wet but the second and third weeks of January were cold with wintry precipitation. There was considerable snowfall in South Wales and the West Country on the 10th with water equivalent exceeding 2 inches at many places in Devon. The last few days were very mild and wet. Up to three weeks without measurable precipitation occurred in west Scotland, north-west England, north Wales and south-east England during the month.

'A similar weather pattern occurred in February, mild and wet at beginning and end with a wintry spell in mid-month but precipitation amounts were in general much greater than in January. More than three and a half times the average occurred in the Mourne Mountains. Over Northern Ireland generally it was the wettest February since 1923. Nevertheless, for the second successive month a fortnight without rain was reported from north-west Scotland.

'The southern half of Britain was much drier than the north in March with up to three weeks without precipitation in the Vale of York, Trent valley, west Midlands and on the south coast.

'There were two heavy April snow falls; the first on 1st-2nd was particularly heavy in northern England with a water equivalent of 2.55 inches at Todmorden. The second occurred on the 14th-15th in southern England and south-west Wales with level depths of 3 inches to 6 inches. The first three weeks were predominantly wet and unsettled in the southern half of the country but less so in Scotland where monthly totals of less than ¼ inch were reported in the Laich o' Moray. It was the wettest April over England and Wales generally since 1920.

'May was generally unsettled and wet but short-lived warm dry spells were experienced. The first of these was continued from the end of April and lasted until about May 3. Possibly the pleasantest spell of the summer at many places started about May 26 and lasted throughout the first week in June in the south-east.

'The rest of June, however, was wet and unsettled in most parts of the country with rainfall often thundery in nature. Among the days of heaviest rainfall were the 9th, when more than 2 inches was recorded over a large part of Down and Armagh with amounts approaching 4 inches in the Mourne Mountains; the 22nd in the Fen District; the 23rd in Londonderry and the Aberdeen-Banff-Moray area.

'Heavy rainfall, with totals more than 2 inches, was experienced in the Home Counties on July 5 and in Pembrokeshire on the 31st where the rainfall was associated with gales of 80 m.p.h. A feature of the period 19th–21st was the dull, cold weather in south-east England (maximum temperature 15°C) contrasted with the sunny warmth of south-west Scotland (maximum temperature 28°C).

'Some of the most spectacular storms of the year were experienced in August. The first of these occurred on the 3rd-4th in eastern Scotland from Berwickshire to Banffshire: more than 3 inches was recorded in the Lothians and Berwickshire on the 3rd, and in Aberdeenshire and Banffshire on the 4th. Over a very extensive area which included the Lake District, Isle of Man, northern Pennines, the Border and southern Scotland, falls of more than 2 inches were reported on the 13th: the heaviest recorded daily fall was 4.20 inches in the Daer valley of south Lanarkshire but eye witness accounts and flood damage suggest that an extraordinary unrecorded rainfall may have occurred on the hills between

Borrowdale and Langdale. Daily falls approached 3½ inches in thundery rain on the 20th in Staffordshire and at Harescombe Grange, in Gloucestershire, the most intense short period fall reported in 1966, 3.30 inches in 90 minutes, was recorded. The following day, the 21st, more than 3 inches occurred in Hertfordshire and Wiltshire, and on the 29th 2 inches was exceeded over a wide area from Hampshire to Leicestershire with 4.15 inches reported at Wroxall, Warwickshire. The period April-August, 1966, was the wettest such period over England and Wales generally since 1931.

'The Lake District suffered further flooding following heavy rainfall on September 3rd when more than 4.50 inches was recorded at Wastwater. The most notable feature of September weather, however, was the long dry period in the second half of the month when most counties in England and Wales experienced a period of more than a fortnight without rain, the last extended rain-free period of the year.

'October was a mainly unsettled month with heavy rainfall for the first three weeks. In Dorset, one-fifth of the annual average was recorded in the 10 days, 13th-22nd, with more then 2 inches in southern Hampshire and Dorset on the 22nd. Somewhat drier weather prevailed generally from the 23rd but extreme eastern Lincolnshire and Kent were marked by curiously persistent and heavy showers coming in from the North Sea in the last five days of the month: nearly 4½ inches was recorded in the Broadstairs area in this period.

November was a month of mixed weather, predominantly cold with northerly winds and wintry showers but with milder periods interspersed. The most notable rainfall of the month occurred on the 4th when more than 2 inches was recorded over a wide area from Herefordshire to Dorset (and also in Down and Armagh). Extensive flooding followed the rainfall in the Bristol Avon catchment area.

'December was unsettled throughout. On the 1st one of the deepest depressions to have affected the British Isles in recent years brought gales and very heavy rainfall to most districts: amounts exceeded 2 inches in the mountains of Wales and in the Lake District. Other days of heavy rainfall were: the 9th with almost 5 inches in the Plynlimon area; the 12th, particularly wet over Exmoor (3.34 inches at Exford) and Cardiganshire; the 17th, in the Lake District and western Scotland; the 19th, again in the Lake District.) 3.67 inches at

Coniston) and the 28th and 30th in Devonshire and on the Brecon Beacons. In all these areas at least 2 inches was recorded on the days named. Special mention should be made of the rainfall at Glen Etive, Argyllshire, on the 16th-17th: 2.26 inches was recorded on the 16th and 7.85 on the 17th, a total of 10.10 inches in the two days. The amount for the 17th is the second highest daily rainfall on record for Scotland. At some time or other in the month flooding affected most of the counties in Britain: the Glasgow area, particularly Paisley and north-west Scotland, was badly affected on 17th-18th. It was the wettest December since 1929 over Northern Ireland and Scotland generally.'

J. GRINDLEY

1967 Although January, apart from the first ten days, was exceptionally mild and dry, little work was done in the fields, owing to the wet soil. Cultivations remained behind schedule. The first fortnight of February brought improved conditions, which enabled field work to go forward, but then wet and mild weather set in again. March brought a welcome respite from the rain, and excellent progress was made with the sowing under almost ideal conditions. Autumn-sown crops proved to have weathered the winter very well and looked promising. There was a good spring bite of grass on light soils, and the lamb crop was on the whole better than usual, the ewes being in good condition. April and May were very variable months, with mild and cold spells alternately, with some sunny and some very wet periods, and with several widespread and severe frosts, which did considerable damage. In consequence the growth of spring crops was rather slow, and some damage from disease and competition with weeds was reported. There was a good growth of grass towards the end of the month, and a fine fortnight in June enabled some excellent early cuts of hay to be taken. Later in the month rain interrupted haymaking, but the weather improved again in July, and on the whole an excellent hay harvest was gathered, both quantity and quality being better than usual. Some damage was done to the cereal harvest by thunderstorms in July in certain districts, though other regions escaped. After changeable weather in the first half of August a fine spell late in the month enabled harvest to make a good beginning, and in the south the barley harvest was nearly finished by the end of the month. The later harvest, particularly in the north, was made difficult by periods of heavy rain in September, but, in general, yields were high and the corn was of fair quality. The autumn rainfall, in October, November and December, was greater than usual, creating difficulties for the harvesting of potatoes, sugar beet and roots and delaying autumn cultivations. However, thanks to the early harvest in the south, autumn sowing was well ahead, particularly on the light lands. Livestock entered the winter in good condition, and there were adequate supplies of winter keep.

Note: This was the autumn of the great foot-and-mouth disease epidemic, which raged through Shropshire and the West Midlands and North, taking a toll of over 200,000 cattle, over 100,000 sheep and over 100,000 pigs, with a total cost reckoned to be in excess of £100,000,000.

RAINFALL IN 1967 GREATER THAN AVERAGE FOR THE THIRD CONSECUTIVE YEAR

Reproduced by kind permission of The Times

'For the third consecutive year rainfall over England, Wales, and Northern Ireland was greater than average. England and Northern Ireland were drier in 1967 than in 1966, and Scotland was about as wet, but Wales was significantly wetter.

'The main features of the year were a very wet March over Scotland, a very wet May, the wettest for nearly 200 years over England and Wales, a dry summer (June to August), and a wet autumn, October being particularly wet. As in several recent years there were no extended spells without rain, the longest dry period of the year, about three weeks, was in June.

'The winter of 1966–7 was generally open and little difficulty was experienced with snow and ice: most of the snow melted fairly quickly in lowland areas.

'Perhaps the most severe short period rainstorm of the year occurred in Lancashire on August 8th, when the village of Wray in the Lune valley was devastated. North-west England, like the south-west peninsula, seems peculiarly susceptible to heavy thundery rain. There was a similar occurrence in August, 1966, at the head of Borrowdale.

'As in the Borrowdale case, it is unlikely that the point of maximum rainfall has been ascertained in the Lune storm. Although the United Kingdom has one of the densest rain-gauge networks in the world (averageing 1 gauge to 16 sq. miles), all too frequently the heaviest rainfall in a storm is not measured and storms may even be missed completely, particularly in hilly districts.

'River flooding, in general, appears to have been more frequent than usual, perhaps because of a late start to the drying out of soils in summer (after the wet May) and the heavy autumn rains.

'The largest totals were probably near 200 inches at Sca Fell. Among the smallest totals were 17·44 inches at Ponders Bridge, Cambridgeshire.

'The weather of January and February followed a broadly similar pattern, the first half of each month being mainly cold and dry and the second half mild and wet. However, precipitation was generally much heavier in February.

'Scattered snow showers carried by northerly winds were fairly general in northern districts about January 5th–8th. Although few daily rainfalls exceeding 2 inches were reported, one such occurred at Hamnavoe (2.49 inches on January 29), an unusual event for Shetland.

'After a mainly dry period stormy wet weather set in about February 15th. The 22nd and 27th were particularly wild and on each of these days rainfall exceeded 4 inches at Cwm Dyli (Snowdon). On each of the days March 2nd to 5th rainfall exceeded 2 inches in north-west counties from Argyll-shire to Sutherland and the 10th, 24th and 26th also were very wet in these districts. On the 24th more than 4 inches was recorded at Kinlochourn, Inverness-shire. From Loch Broom to near Loch Tay four times the March average was recorded.

'It was the wettest March since 1903 over Scotland generally. Over southern Britain precipitation was mainly restricted to the 7th to 12th, and to a spell beginning on the 25th which became wintry on Easter Day.

'Most of the April precipitation occurred on the 1st-2nd and in spells between the 5th-10th and 20th-24th, but exceptionally heavy rainfall was infrequent even in the hills. Periods of two weeks or more without rain occurred widely in

Wales, around the Solway Firth, and in the Isle of Man from about the 3rd.

'May began most unpromisingly with snow as far south as London on the 1st and widespread damaging frosts on the next two nights. The month was unsettled and wet throughout, with thunderstorms nearly every day in one part of the country or another. The 11th was an extremely wet day in Manchester, where 3.38 inches fell in about six hours. It was the wettest May since 1773 over England and Wales generally.

'June provided a welcome contrast. Over England and Wales most of the rain fell in a thundery spell about the 23rd–25th and many places in the south recorded more than three weeks without rain from May 30th.

'July was mainly dry at first, particularly over England and Wales, but this dry period was brought to an end on the 13th when thundery rain was widespread. Some of the heaviest daily falls of the year were recorded in Northern Ireland where 3.35 inches occurred at Annalong Valley in the Mourne Mountains. At Oxford 2.44 inches fell in 35 minutes on the 13th. Further violent thunderstorms were reported in south-east England from about midnight to 6 a.m. on the 23rd; the highest amount recorded, 3.93 inches was at West Byfleet, Surrey. On the 29th more than 4 inches occurred in Ogmore Vale.

'In the heavy rainfall of August 8th, which affected mainly the Bowland Forest and adjacent Lune valley, authentic falls of more than 3 inches were recorded in less than two hours and it is estimated that in the middle of the Forest of Bowland about 4.60 inches fell in about 90 minutes. Further violent thunderstorms were reported in the Nottingham area on the 10th, when the heaviest fall, 2.66 inches, occurred in about an hour at Welford. Much more settled, drier weather set in about the 19th and persisted until near the end of the month.

'The last day of August marked the beginning of a long unsettled spell, which lasted with but few breaks until well into November. September 18 was a very wet day in the Mourne Mountains, and on the 20th a heavy fall of 3.46 inches in the Wirral caused flooding.

'October was unsettled throughout, with frequent heavy and prolonged rain, which gave rise to floods in many parts of Britain as the focal point of the rainfall shifted from day to day. On the 1st, 4.2 inches occurred at Bryn Gwynant, Caernarvonshire, most of the rain falling in 11 hours. On the 3rd, the Lake District and Wales and on the 6th the Lake District and central and south-west Scotland were affected.

'The 8th was a particularly wet day in the Lake District and in counties bordering the Solway Firth. At Great Langdale, Westmoreland, 3.04 inches fell on the 8th and 3.01 inches on the 9th, but the observer estimates that about 6 inches fell within 24 hours. The day of heaviest rainfall generally was the 16th, when a general depth of 1.2 inches (weighing 45,000m. tons) is estimated to have fallen over England and Wales. Daily falls exceeding 1 inch over the country generally are most unusual. The heaviest daily fall of the year (in the standard rainfall day) occurred on the 16th when 5.48 inches was recorded in the Brecon Beacons.

'The unsettled weather continued unabated in November culminating in a very wet Guy Fawkes Day. Rainfall amounts exceeded 2 inches in parts of southern England, in Ross and Cromarty and widely in Antrim and Down on the 1st. Amounts for the 4th–5th exceeded 4 inches in the upper Tees valley. A much more settled, drier spell intervened from 15th to 24th before weather again became disturbed in the last week of the month.

'Precipitation in December occurred mainly in a snowy spell from the 6th to 11th and in a mild spell from the 18th to the 25th. Snowfall was heavy on the 8th and 9th in the north and west of the country, and on the South Coast, where transport was disrupted. Level depths of show of 6–8 inches were widely reported in the West Country, Wales and on the Pennines. The snow soon melted as mild air with rain and drizzle moved south on the 11th. The weather became colder after Christmas, and the year finished with wintry showers.'

J. GRINDLEY

1968 Most autumn-sown crops came through the winter in good condition, with some local damage by slugs and leather-jackets. The weather in late February and March was ideal for spring cultivations, and as a result work was more forward than usual. The lambing season proved better than average, and winter keep was adequate. The dry, cold weather in early April, however, delayed the growth of spring grass, though rapid growth was made later in the month. The dull, cold weather in May gave a setback to some of the spring sown corn, but on the whole prospects during spring were good. An excellent first cut of grass for silage was taken, and the early hay was likewise got in in good condition.

The real trouble began in July when severe storms, including many thunderstorms, seriously damaged cereal crops, particularly barley. A heavy hailstorm in Yorkshire did much severe damage and other hailstorms, combined with floods, ruined crops in Devon. The main hay harvest proved very difficult, with much hay ruined by rain and flooding. In August the situation steadily deteriorated, with many crops being badly flattened and twisted by rain and wind, while heavy growth of weeds hindered harvesting. The quality of the grain proved to be poor. More periods of very heavy rain in September made things even worse. Both quantity and quality of the cereal crops suffered. Lifting potatoes was difficult, and virus yellows had appeared in sugar beet. Beans were proving a very late harvest and were being attacked by chocolate spot. While the pastures had plenty of lush grass, the feeding quality was poor and lambs and beef cattle were slow to finish. Hay and straw of good quality were scarce.

Autumn proved to be almost as unfavourable as summer, with much rain and dull weather. Harvest dragged on into November in many districts, and autumn cultivations were in general very slow and difficult. Fewer crops of autumn cereals than usual were sown. On the other hand grass was still abundant, and roots and kale crops were giving a heavy yield.

1968 SUMMER WAS WETTEST SINCE 1931

Reproduced by kind permission of The Times

'For the fourth consecutive year general rainfall over England and Wales has been greater than average and in each year the margin has been quite substantial. The period 1965–8 is the wettest period of four consecutive calendar years over England and Wales since 1927–30. In each of the years 1965 to 1968 the excess rainfall has been mainly attributable to the summer six months (April–September). Total rainfall for these months in 1968 over England and Wales generally was the greatest since the summer of 1931.

'Scotland and Northern Ireland, on the other hand, had a generally dry year. One of the outstanding features of the weather of 1968 was the contrast between the mainly fine, dry summer in north-west Britain, including North Wales, north-west England, Northern Ireland and particularly the extreme north-west of Scotland, and the depressingly dull, cool, wet summer in south-east England, the east Midlands and East Anglia.

'The year has been remarkable for a series of notable widespread rainstorms. Outstanding events included March 25–26, when recorded rainfall approached 10 inches at places in the North-West Highlands of Scotland; July 10th, when south-west England, the south and east Midlands and Lincolnshire suffered from a damaging storm which produced amounts exceeding 6½ inches to the east of Bristol.

'On September 14th–15th it was the turn of south-east England and East Anglia to suffer extensive flood damage from comparable and perhaps rather more severe falls, and on September 20th–24th amounts approaching 10 inches were recorded in the southern Pennines. On October 31 more than 6 inches was recorded in Co. Down. Both the falls of July 10th and September 14th–15th were quite outstanding for the extent of very heavy rainfall over lowland Britain.

'The largest rainfall total was probably 162 inches on Sca Fell. The smallest total so far noted was 19.23 inches at Manningtree, Essex.

'There was a dry start to the year, precipitation for January and February being well below average for the country generally. The distribution of precipitation was somewhat similar in both months, the first half of each month being predominantly wet and the second half dry. Heavy snowfall occurred in Scotland on January 4th–5th and a blizzard was experienced in parts of Wales, midland and southern England on the 8th. Flooding occurred in south-west England, where heavy rain rather than snow fell on the 8th. Stormy, milder weather brought a thaw to most parts of the country on the 14th, and more general flooding occurred.

'Snowfall was heavy in Scotland and northern England on February 4th–7th, in parts of Cheshire the water equivalent of the snowfall on the 5th exceeded 2 inches.

'A long mainly dry spell began about February 14th and lasted to about March 12th. Most parts of England, Wales and southern Scotland experienced at least a fortnight with negligible precipitation, and in south Wales and south-west England a month elapsed with virtually no rain.

'The period from March 12th to 14th was unsettled and wet, particularly in northern England and Wales. Much snow fell on the 22nd, and its thaw in heavy rain on the 23rd may have contributed to the extremely high floods in the Eden Valley on March 23rd–24th and the Vale of York on the 25th.

'The storm of March 26th–28th in north-west Scotland has already been noted. This storm in many respects eclipsed the previous major storm of December 16th–17th, 1966. The inferred extent of areas with more than 10 inches exceeds by a wide margin any earlier estimate of areas covered by more than 10 inches in a two-day storm.

'Heavy precipitation occurred on April 1st in the Lake District and Forest of Bowland, but from the 4th to the 14th weather was dry nearly everywhere. A warm thundery spell began about the 16th, and storms were particularly intense on the 17th–18th and 20th.

'Heavy rainfall accompanied by flooding occurred in southern and eastern Scotland on May 4th–5th and the period up to the 13th was generally wet. A more settled spell began about the 14th and continued to the end of the month in most places. Scattered outbreaks of heavy rain occurred, however, in southern and midland England on May 24th–27th.

'The pattern of generally drier but not completely rainless weather established in mid-May was maintained with isolated and occasionally heavy thunderstorms up to June 18th. In one of the thunderstorms on the 3rd at Prickwillow, Cambridgeshire, 3.26 inches was recorded in 140 minutes. The period from the 18th to the 28th was extremely unsettled and wet over southern Britain.

'The weather was hot at the beginning of July. Multicoloured dust, thought to be of North African origin, was brought down in many areas by rain. Heavy thunderstorms occurred on July 1st and 2nd in Wales, northern England, southern Scotland, and Northern Ireland. Severe hailstorms associated with these storms did a great deal of damage to crops. At West Baldwin reservoir, Isle of Man, 5.83 inches was recorded on the 2nd.

'The extraordinary rainfall of July 10th was associated with warm and cold fronts and a narrow warm sector which brought very moist air into southern England. The track of the depression was from Land's End to the Wash, and it was along this track that the heaviest rainfall occurred. Apart from the area of heaviest rainfall to the east of Bristol, quite extensive areas with more than 4 inches were apparent from the Quantock Hills to the Lincolnshire wolds. Extensive flooding occurred along the line of heaviest rainfall.

'The period from July 17th to August 12th was mainly warm and sunny over much of western and northern Britain, where a spell of up to three weeks with virtually no rainfall was recorded. Eastern, midland and south-eastern England, on the other hand, retained almost unbroken cloud cover and periods of heavy rainfall occurred, particularly on August 2nd and 6th to 9th. From August 12th to 19th, northern England and southern Scotland began to experience heavier rainfall, and the 19th was particularly wet in the mountains of North Wales, where more than 5 inches was recorded.

'The outstanding weather event of September was undoubtedly the heavy rainfall of the 14th–15th. The line of heaviest rainfall extended from the Hampshire Downs, a little to the south of the North Downs, across the Thames estuary and into East Anglia. Peak amounts for the two days were 7½ inches in the Edenbridge area and more than 8 inches at Northchapel, Sussex. Most of the rain fell in a period of 15 to 18 hours from midnight on the 14th to the afternoon of the 15th.

'The storm had been preceded on September 11th–12th by very heavy rainfall extending from west Norfolk to south-east Scotland; amounts approached 4 inches on the 12th in the Borders. North-west Britain meanwhile had been experiencing fine dry weather, but this spell was brought to an end in North Wales and north-west England on the 20th by four days of very heavy rainfall; in this period as much as 10 inches was recorded in the southern Pennines. It was the wettest September over England and Wales generally since 1918.

'Heavy rain fell in northern England and southern Scotland on October 1st, with amounts exceeding 4 inches in the Lake District. In most places, however, the first week of October was mainly dry. Rainfall was generally heavy from the 7th to the 12th, but amounts became progressively less thereafter over much of England and Wales; parts of northern England recorded little or no rainfall from the 12th to the 26th. The last few days of the month were much more unsettled, and the 30th and 31st in particular were very wet: at Tollymore Park, Co. Down, 6.26 inches was recorded on the 31st, and this was followed on November 1st by a further 3·19 inches.

'Rainfall was heavy on November 1st, particularly in the east Midlands, with amounts exceeding 3 inches to the east of Louth, but thereafter a long mainly dry spell persisted up to about the 19th. The last 10 days of the month were much more unsettled.

'The first half of December was mainly dry in most parts of the country, although south-west England experienced heavy thundery rain on the 9th and heavy rainfall occurred in north-west Scotland on the 12th–13th. The second half of December was more unsettled, with heavy rain in the south and snow in the north of Britain.

'By the 24th, snow had spread to the whole of eastern Britain and amounts were quite heavy in many places. The precipitation occurred as rain in south-west England, where values exceeded 2 inches and flooding occurred. From the 25th, winds were northerly and there were some moderate falls of snow, with extensive drifting in the east.

J. GRINDLEY

1969 A salient feature of 1969 was the wet spring, made all the more serious because of the heavy rains in the autumn of 1968, which had resulted in less winter wheat than usual being drilled. Spring sowing was so delayed that some fields were never planted, while other crops went in under such bad conditions that yields suffered severely. The total cereal acreage was down by some 300,000, but where sowing had been completed in reasonable time the harvest turned out to be quite good.

Grass also suffered in the inclement spring but later recovered and produced an abundance of autumn grazing. Also owing to the weather, the potato acreage failed by 36,000 to reach the target figure of 650,000 acres, leaving a deficit in potato supplies which had to be made up by considerable imports. Sugar beet yields were below average.

The national cattle herd increased by some 225,000 to attain the record total of 12,374,000 animals. It would doubtless have been higher but for the fact that farmers in the Midlands were replenishing their herds after the disastrous losses through foot-and-mouth disease in the previous year. The greatest increase was in beef cattle, and at the end of the year the numbers of dairy cows were showing a downward trend.

The decline in sheep numbers, which had begun in 1965, continued, with a decrease of about 2,000,000 over 1968. The adverse spring resulted in heavy lambing losses. Also, changes in methods of husbandry on lowland farms were reducing the markets available to store lambs from the hill farms. Pigs, however, were on the upgrade, reaching levels previously attained only in 1965. The number of pigs coming forward for slaughter was 6% higher than in the preceding year, and the breeding herd showed a similar marked increase. Poultry numbers and the output of eggs remained steady, matching almost exactly the national demand, though producers complained that the increasing costs of food and overheads were reducing their profits.

The Government increased the subsidy on fertilizers and lime and announced that compulsory projects for the eradication of brucellosis would begin in 1970.

RAINFALL UP AGAIN IN ENGLAND

Reproduced by kind permission of The Times

'For the fifth consecutive year rainfall over England and Wales generally was greater than average in 1969, but the excess was small and in many respects the year was far more agreeable than its four predecessors. Over Scotland and Northern Ireland general rainfall was very much less than average; it was the driest year since 1955 over Scotland.

'There was a marked contrast between the incidence of rainfall in the first part of the year over England and Wales generally and that over Scotland. Over England and Wales rainfall for the first five months was 41 per cent more than average; there has been no wetter start to a year since 1951. By the end of May, farmers in many parts of the country, particularly the Midlands, despaired of getting spring sowing completed.

'In Scotland, on the other hand, the period January to April was 21 per cent less than average, the driest such period since 1956.

'Over England and Wales generally, rainfall for the period June to October was only 68 per cent of average; in the past 100 years there have been only three drier such periods, and these occurred in the memorable dry years of 1921, 1947 and 1959.

'Over Northern Ireland there has been no drier period June to October since comparable records began in 1900. The autumn was outstandingly dry. Over England and Wales September–October combined was the driest such period since 1752 (when however September was 11 days short).

'Although the dry weather was welcome to most people, it brought its own problems, particularly to water undertakings in south Wales, north-west England (including the central Pennines), south-west Scotland and Northern Ireland, and to farmers in eastern Britain who found it difficult to proceed with autumn cultivation because of rock-hard ground.

'Recent years have been notable for widespread, major rainstorms which have inundated considerable areas; and 1969, too, provided an outstanding example. The storm occurred on July 28th and affected mainly south-west England, south Wales and the Midlands. The storm ranked second only to that of September 16th, 1967, over the Midlands and northern England in respect of area covered by amounts of 1 inch or more, and was without precedent in respect of areas covered by falls of three and four inches or more.

'Although flood damage was extensive, the severity and cost did not compare with that arising from the storms of July and September, 1968, possibly because there was no very intense fall (more than 6 inches in a day) and the ground was in a much better state to absorb water than in either of the 1968 storms.

'The largest rainfall total was 133.80 inches at Styhead Tarn, Cumberland. The smallest total so far noted was 18.16 inches at Erith, Kent.

'January was generally a mild and open month with frequent depressions crossing the country, particularly in mid-month. The wettest day generally over the country was the 20th when amounts exceeded 2 inches over a wide area of

Wales, north-west England and south-west Scotland, with more than $4\frac{1}{2}$ inches in Snowdonia.

February, by contrast, was predominantly a cold and snowy month with periods of heaviest snowfall about the 6th–7th, 10th–11th, and 19th–20th. The snowfall of the 19th was immediately followed in southern England by heavy rainfall and a general thaw which brought widespread flooding.

'March continued the predominantly cold weather of February, most of the precipitation occurring in the period between the 10th–19th. The 12th was a day of particularly heavy precipitation with blizzard conditions in the east. Extensive flooding came after the melting of the snow in the Midlands and East Anglia on the 13th. Most stations had long dry periods between the 1st and 9th and the 20th–28th.

'The first eight days of April were dry everywhere and the Easter period was particularly sunny. Precipitation occurred mainly in two spells from the 9th to the 15th and the 21st to the 27th.

'May was an exceptionally wet month generally. Precipitation occurred on most days and was often heavy and thundery. The Midlands were particularly hard hit, heavy falls occurring at one point or another on the 5th, 16th–17th, 24th–25th, 27th and 29th. Amounts exceeded 2 inches at a number of places in Staffordshire on the 24th and in east Wales and the west Midlands on the 25th. Nevertheless, parts of north-west Scotland remained dry, Fort William having its driest May for 10 years.

'June 2nd was wet in northern England and southern Scotland but the period 3rd–13th was one of the driest and warmest of the summer over the country generally. The fine weather was ended by severe thunderstorms in the Midlands and northern England; at Worsborough, near Barnsley, 2·33 inches fell in 45 minutes, an exceptionally intense fall. It was reported that barrowloads of melting hail were swept up the next day. The period up to the 23rd remained very unsettled.

'The mainly dry weather which began about June 24th continued over England and Wales to near the end of July but a notable break occurred on the 6th in south-east England when more than 1 inch was recorded over an area south-east of a line from Portland Bill to Orford Ness. The greatest rainfall was recorded in the Ramsgate area where amounts approached 3½ inches.

'The heaviest rainfall in the storm of July 28th occurred along a line from Plymouth through the Midlands to Flamborough Head. The area hardest hit was south-west England where more than 4 inches was recorded over a wide area extending from Wendron Moors to the Mendip Hills. The highest daily total, 5.71 inches, was reported at Ellbridge, near the Tamar estuary.

'Very heavy rainfall occurred also in Kent and the east of East Anglia; amounts exceeded 2 inches in eastern Kent. The 29th was a wet day in the east Midlands and western parts of East Anglia where totals approached 2 inches. Rainfall in August was mainly scattered and thundery.

'A notable storm occurred on the 2nd along a track from Kent to the west of London and across the Midlands into the Wirral; more than 4 inches fell at Kempton, Middlesex and in Warwickshire. Parts of Devon were flooded after intense storms on the 8th. These storms extended to the west Midlands and were remarkable for the intensity of the lightning and violence of the thunder.

'On the 11th–12th it was the turn of south-west England, parts of Wales, northern England and the north Midlands to have severe storms; more than 3½ inches was recorded in the Brecon Beacons.

'At Manston, east Kent, more than 4½ inches occurred on the 14th–15th, including a fall of 1.93 inches in an hour. At Coulter Reservoir, south Lanarkshire, 3.06 inches fell in two hours in a local storm on the 16th; a fall of this intensity is extremely rare at a Scottish station.

'The first nine days of September were dry and settled over most parts of the country except north-west Scotland. Heavy rainfall occurred in the Lake District on the 9th (when 4 inches was recorded in Ennerdale) and again on the 10th when south Wales and Devon were affected.

'Heavy thunderstorms were reported in southern and south-west England on the 18th; more than 2½ inches was recorded at Taunton. The remainder of the month was mainly dry over England and Wales but heavy rainfall occurred in north-west Scotland. Rainfall was almost nil in parts of west Suffolk and Norfolk for the whole of September.

'Although precipitation in October occurred on several

days, amounts were slight over much of England. The north-west of the country was much more unsettled, however, and amounts exceeded 2 inches in one part or another of north-west Scotland on the 8th, 13th, 15th–16th, 22nd–23rd and 30th.

'The warm, dry weather of October over England and Wales continued for a few days in November but thereafter it became thoroughly unsettled and wet for a fortnight over the whole country.

'Other exceptionally wet days over England and Wales were the 11th, 14th and 16th. The wettest day of the year in Northern Ireland occurred on the 21st when rainfall exceeded 2 inches widely over the province; locally values exceeded 4 inches, the highest amount being 5·43 inches at Ballybraddin Forest, Antrim.

'The 21st was a very wet day in the central lowlands of Scotland. Weather became drier and colder from the 18th and from the 24th very much colder with frequent snowfall. A feature of the month was the unusual number of thunder-storms which were often associated with snowfall.

'In December, precipitation occurred generally on most days in the first three weeks and was particularly heavy in north Wales and the Lake District on the 13th when more than 3 inches was recorded. Snowfall occurred in the periods 3rd–4th, and 16th–19th but in general amounts were slight. The last week was drier but the weather became very cold and slight snowfall occurred in the last few days.'

J. GRINDLEY

1970 1970 can be classified as one of the better farming years. Winter wheat was the crop of the year, yields in most districts being well above average. Barley, on the other hand, gave poor results, being badly affected by a spring drought and later by foliar diseases. The same drought played havoc with spring grass. However, there were no spring frosts.

In summer heavy rains in July spoiled much hay, but they started the grass and roots growing again after the drought. Harvest was characterised by sunshine without drought, a pleasant combination. Then a mild and quiet autumn allowed ploughing, sowing and other work to get well ahead, and much autumn corn was sown, often by farmers in such a hurry that they neglected to give the soil a proper cleaning.

The year ended with moderate supplies of hay, silage and roots in store, though without any great surplus. Straw was in short supply and consequently dear.

Although weatherwise the year was satisfactory, for farmers in general it was difficult because of economic pressures. Loans continued to be expensive, and increases in such overheads as electricity, water, postage and telephone, and in such essentials as fertilizers, fuel and labour, accelerated. Farmers found they needed more and more units, such as larger numbers of acres, cows or hens in battery cages, or else a higher return per unit, such as higher yields of milk per cow or more grain per acre, in order to maintain their income.

The intensification thus demanded was being increasingly pursued without due regard to the basic principles of husbandry. Summer fallowing became the exception rather than the rule, and continuous cereal growing over a series of years with never a break was now recognised practice. Men who had once considered they were doing good work if they milked 10 cows now found themselves coping, single-handed, with a herd of 100 or 120. Similar developments were occurring in the pig and poultry industries.

The total wheat acreage was a record, at 2,495,000 acres, but the barley acreage showed a small decline, and the total cereal harvest was down by about 500,000 tons. The potato acreage exceeded the target by some 19,000, resulting in a surplus tonnage and low prices.

The national cattle herd increased, largely through an expansion of about 100,000 in the breeding herd of beef cows. There was a marginal fall in numbers of dairy cows, through drastic culling in view of the limited fodder supplies. Sheep continued to decrease in numbers, though this year the decline was small (only about 400,000). Pig breeders were evidently optimistic and expanded their breeding herd by some 20,000 sows, to the highest total (774,000) of recent years. Numbers of poultry increased very considerably, especially in the table poultry sector.

In this year several Continental breeds of cattle, notably the Simmental and Limousin, established themselves in Britain.

SIX-YEAR PERIOD 1965–70 THE WETTEST ON RECORD SINCE 1927–32

Reproduced by kind permission of The Times

'Rainfall last year was slightly greater than average over Wales, Scotland and Northern Ireland and precisely the average for England. The general value for England and Wales combined was almost the same as that for 1969 and, although the excess over the average was small for both years, the series of consecutive years with rainfall greater than average is six. The series of wet years 1965–1970 was the wettest period of six consecutive years since that for 1927–1932; it is worth recording that the year following that series, 1933, was one of the driest on record over England and Wales.

'Notable features of the rainfall distribution throughout the year were the wet first four months, particularly over England, Wales and Northern Ireland; the long series of dry months over England and Wales extending from May to October and the very wet November over the whole country.

'The total rainfall for January to April 1970 was the greatest for such a period since 1951 over England and Wales. The summer was dry; the total for May and June combined was the lowest for these two months since 1940 and in recent years only the memorable summers of 1959 and 1964 have had lower totals for the period May to October. Although September and October were dry, pleasant months over England and Wales, the combined totals were not nearly as low as those for 1969. The period July–November was the wettest over Scotland generally since the outstanding wet year of 1954. The Northern Ireland December value was the second lowest in this century.

'The largest rainfall total was 195.70 inches at station Delta on the slopes of Snowdon. The smallest total so far noted was 18.43 inches at Burnham, Essex.

'The first week of January was cold with heavy snowfall in the north, particularly on the 5th–6th. A belt of snow crossed southern districts on the 8th, but a thaw quickly followed and from the 9th the weather was mainly mild and wet; frequent depressions crossing the country brought rain on most days. February started mild with heavy rain, particularly in north-west Scotland, but weather became progressively colder from the 3rd as a northerly airstream became established. Precipitation in the period 6th–17th occurred mainly as snow, the most notable occasions being on the 12th when a blizzard swept southern Britain and the 17th when snowfall was widespread over the country.

'Weather became milder and very wet from the 19th but cold northerlies with snow at times were reestablished from about the 25th. This cold weather was continued in March and precipitation, which was frequent, occurred almost entirely as snow up to the 13th. The snow fell mainly in showers, but on the 3rd–4th a small depression moved from Northern Ireland to Kent bringing blizzard conditions along its track. Level depths of snow of more than 12 inches were reported in the Midlands and south-east England. A spell of westerly winds occurred from the 16th–22nd and rainfall was heavy in north Scotland on the 16th, but northerly winds returned from the 26th bringing snow showers and a cold Easter.

'The cold weather of late March continued well into April and snow showers in a northerly airstream were frequent up to the 10th. Blizzards occurred in northern England on the 1st and snow was widespread on the 8th. Very heavy rainfall occurred over the southern Pennines on the 12th–13th. Milder, westerly weather brought wet conditions to hilly western Britain in the period 14th–23rd. The period 21st–23rd was particularly wet and on the 22nd, 7·15 inches was recorded at Seathwaite, Cumberland, almost certainly the largest amount for an April day on record in the United Kingdom. Colder weather returned on the 26th with more snow showers in the north.

'Most of the precipitation in May occurred in the second week over England and Wales and was generally thundery in character. Heavy thunderstorms occurred in East Anglia and Oxfordshire on the 28th. Scotland and Northern Ireland experienced less settled weather but the month was by no means wet generally in these countries. May saw the beginning of the longest spell of dry weather in the year. Within the period May 11th to June 22nd, many places in Britain experienced up to three weeks without measurable precipitation. The first week in June in particular being fine and hot. In parts of the east Midlands and East Anglia as many as 42 days without rain were recorded. In spite of the generally dry weather, June was notable for violent local thunderstorms and in particular for the number of rainfalls classified as 'very rare' (falls defined as recurring at any one point once in 160 years or even less frequently).

'The thundery outbreak began in Scotland about the 6th and spread to the rest of the United Kingdom in the following days, the storms being particularly widespread on the 10th–11th. Among the more outstanding falls were: 3.65 inches in 117 minutes at Lossiemouth on the 10th, 4.37 inches in 90 minutes at Miserden, Gloucestershire, on the 10th, 2.64 inches in 25 minutes at Pershore on the 11th, 5.00 inches in three hours at Lochgoilhead, Argyllshire on the 11th. Weather became generally unsettled from June 22nd and further severe thunderstorms occurred in the east Midlands and East Anglia on the 27th; the most notable of these falls was that at Wisbech when 2.00 inches fell in 10 minutes, almost certainly a United Kingdom record for such a duration.

'The first week of July was cool but mainly rather dry. Temperatures later became very high and on the 7th–8th there were isolated thundery outbreaks. The unsettled weather continued until about the 28th when a drier warmer spell began. Some heavy rainfall was experienced in hilly areas of Scotland, particularly on the 5th (4.83 inches at Ardgour, Argyllshire), the 6th, 11th–12th, 24th in north-east Scotland where the rainfall was thundery in character (4.53 inches at Netherley, Kincardineshire) and on the 27th in the Tweed valley.

'The first few days of August maintained the fine warm weather of late July but thunderstorms developed about the 5th–8th. In central London 1.69 inches fell in 30 minutes on the 7th, disrupting evening traffic. Northern Ireland experienced its heaviest rainfall of the year on the 15th as a depression moved north-eastwards across the province and over Scotland towards the Moray Firth. Rainfall amounted to 3½–4 inches generally in the Belfast area (highest total 4.64 inches at Kilroot, Antrim) and extensive flooding occurred in the city. Flooding was widespread in south-western Scotland where heavy falls were reported on the 15th and in north-east Scotland where the heaviest rainfall occurred on the 16th (3.85 inches at Lochindoris Lodge, Morayshire). Further heavy rainfall was recorded in the south-west of England and

Midlands on the 19th with totals approaching 4 inches at many widely scattered places. This period of heavy general rainfall continued until the 22nd but from the 23rd there was a week of almost rainless weather over the whole country.

'Apart from a short-lived dry spell from the 3rd to the 6th in southern Britain, the first half of September generally was unsettled with showers or periods of rain over the whole country. The unsettled weather continued for most of the month in northern Britain, but a dry, warm spell lasted for almost a fortnight in the southern half of the country. This spell came to an end as general rainfall spread from the west on the 29th.

'A notable feature of the October weather was the contrast between the heavy rainfall, mainly orographic in character, in hilly western districts and the generally dry weather of eastern and southern Britain. Rainfall amounts exceeded 2 inches on 12 days over an area extending from South Wales to extreme north-west Scotland. These falls occurred mainly at the beginning and end of the month; outstanding among the falls was that on the 4th at Dalness, Argyllshire, when 6.02 inches was recorded. A feature of the weather in many parts of England was the series of warm, golden days in mid-month with misty mornings and sunny afternoons. Virtually no rain fell over England and Wales in the week beginning the 12th.

'November began wet and unsettled and continued so almost throughout. There was a short spell of quite foggy weather in the fourth week but from the 27th it was again unsettled and wet. The heaviest recorded daily fall, 4.16 inches occurred on the 23rd at Brotherswater, in Westmorland, and this was a very wet day over southern Scotland as far north as the Central Lowlands. The 14th was a wet day in southern England and East Anglia where amounts exceeded 2 inches in several places; a little snow was associated with the precipitation in East Anglia.

'The unsettled wet weather of November continued for about six days in December but, thereafter, a settled and at times foggy spell began and lasted for about a week in Scotland and up to about the 22nd over most of England and Wales. The most notable feature of the month was the snowfall which began on the 21st. By Christmas Eve the snow had spread to most parts of eastern and central Britain and by the 26th to all but the extreme west. Despite temporary thawings there were further falls of snow up to the 29th and the year ended on a wintry note over most of Britain.'

J. GRINDLEY

1971 1971 followed the pattern of reasonably good harvests and exceptionally favourable autumns. Once again, winter lingered late in spring, and summer late in autumn. However, in spite of chilly, damp weather in spring, there were no disastrous May frosts. Heavy rain in June spoiled many of the early hay cuts, but later crops supplied some compensation. The corn harvest was excellent and gathered without undue difficulty, but sufficient rain fell to keep grass growing right through summer and autumn.

The open autumn allowed cultivations and sowing to progress with little interruption, with the result that a record acreage of winter corn for 1972 was sown.

Wheat had a near record acreage of 2,710,000, and the barley acreage was slightly up, though oats continued to decline. Yields of cereals in general were well above average. Potatoes also produced a bumper harvest, and the sugar beet crop was the biggest ever recorded.

The dairy herd remained fairly constant in numbers, though milk yields showed an upward trend. There was a continued increase in the numbers of breeding beef-type cows, to a record 1,387,000, and more calves than usual were retained for beef. Sheep numbers became more or less stabilised, after the decline which had characterised recent years, and the lambing season was good. Pigs had an up-and-down year, with a quite dramatic rise in the breeding herd in the early months, followed by a decline in numbers in autumn. Poultry numbers were more or less static.

Average net incomes for farms rose by about 9%, but in Scotland by 23%, largely through the favourable conditions for livestock. The Government announced changes in the system of agricultural support which had prevailed for the past two decades.

Decimalisation of the coinage was instituted.

RAINFALL IN 1971 WAS WELL BELOW AVERAGE

Reproduced by kind permission of The Times

'After a six-year period of above average rainfall, the total for

1971 over England and Wales was well below average, the deficit amounting to 4.3 inches or 12 per cent. Years as dry as 1971 may be expected to recur about one in 10. Rainfall over Scotland and Northern Ireland, too, was well below average, the deficits being even more notable than for England and Wales.

'Notable features of the rainfall distribution throughout the year were the dry periods February to May and September to December, and the very wet June over England and Wales, the long series of dry months over Scotland, only March having greater than average rainfall, and the very dry December over the whole United Kingdom.

'The period May to September, 1971, was the driest since that of 1959 over England and Wales and there has been no drier period, September to December, since 1879, although September–December, 1904, was as dry. It was the wettest June since 1958.

'Among the largest and smallest rainfall totals recorded were 141.7 inches at station Delta, on the slopes of Snowdon, and 16.90 inches at Austerfield, near Doncaster.

'The first few days of January were cold, with light snow in many places. A change to milder weather began about the 5th and heavy rainfall occurred, particularly in hilly western districts, on the 6th and 7th. The next week was again mainly dry in most places but from the 16th predominantly wet weather prevailed. Colder weather returned towards the end of the month, with snow in places.

'The 1st to 10th February was dry over most of the country. Precipitation occurred mainly in the period 11th–20th, hilly districts from North Wales to Argyllshire being particularly wet from the 11th to 14th. At Great Langdale, Westmorland, 5.20 inches was recorded on the two days 11th–12th. The month was mainly dry from the 21st, although a little snow fell on the 27th–28th.

'The first week of March was extremely cold in southern Britain and heavy snowfall was recorded on the 1st. Most of the month's precipitation occurred as rain in the week from 13th–19th and also about 23rd–25th.

'The first three weeks of April were mainly dry and many places in midland and northern England reported a fortnight or more without precipitation in the period. Most of the precipitation in April occurred within the period 22nd–24th, when some places in the Midlands and northern England recorded 50 hours of almost continual rainfall. The 23rd was the wettest rainfall day and was almost certainly the wettest day of the year.

'A late heavy snowfall occurred over southern England on the 25th but the snow soon melted and the rest of the month was warm and dry.

'The first three weeks of May, too, were mainly dry over England and Wales, although thundery rain, locally intense, occurred on the 6th to 7th and the 15th was generally wet. Weather became much more unsettled from the 22nd. The mainly dry weather was confined to the first half of the month over Scotland and Northern Ireland. The 23rd was a very wet day in Angus, with falls exceeding 2 inches at several places.

'Apart from isolated thunderstorms in the English midlands on the 2nd, the first week of June was dry nearly everywhere but the month was notable for an extremely wet period from the 8th to 19th. The rainfall excess was particularly marked in southern England and Wales, where on the 10th amounts exceeded 2 inches over an area extending from Bristol to the outskirts of London and down to the south coast.

'The 18th was as wet a day generally as the 10th; values exceeded 2 inches in an area extending from Sussex to south-west Wales. The period 20th–24th was drier over England and Wales but the last week was again unsettled generally.

'Many places in England and Wales experienced three weeks without rain in the period June 29th to July 19th. Scotland and Northern Ireland, too, were mainly dry in this period.

'Widespread flooding and damage were reported in the Flintshire-Denbighshire area where totals approached 4 inches on the 3rd. Most of the rainfall occurred in a relatively short period: for example, 3.19 inches was recorded in 90 minutes at Ruthin. The period from the 20th was more unsettled but many places recorded only slight or moderate amounts of rain.

'Heavy rainfall did occur on the 24th from the English Lake District to the Tweed Valley (3.21 inches was reported at Newcastleton) and thunderstorms were reported over an area extending from the Fylde to the Vale of York on the 23rd and 25th. In extreme east Norfolk, more than 4½ inches is reported to have occurred in the Great Yarmouth area on July 28th and 29th.

'August was a thoroughly unsettled month in the southern half of Britain. Generally, the third week was drier but thunderstorms occurred in the south towards the end of the week. More than 2 inches fell in and around London on the 4th. One of the most intense storms on record for Northern Ireland also occurred on the 4th when 2·50 inches fell in 78 minutes at Londonderry. The heaviest daily fall of the year in Northern Ireland, 4·12 inches, occurred at Ballypatrick.

'Over England and Wales, most of the precipitation in September fell within the period 23rd to 26th. The second extended dry spell of the year occurred from about August 30th to September 22nd and many places in the south experienced up to three weeks without rain in this period.

'In October precipitation over England and Wales was mainly concentrated into the period 12th–19th. The 18th was very wet over Wales, northwest England and the West Riding, where 4 inches was exceeded in the southern Pennines. The period of wet weather was more extended in Scotland but there the first five and the last seven or eight days of the month were mainly dry.

'The first half of November was mainly dry over England and Wales, apart from 5th–8th. In parts of East Anglia, a fortnight without precipitation was recorded between October 20th and November 4th.

'A spell of wintry weather brought snow to most districts from 18th–23rd and many roads were blocked in the northern half of Britain. Precipitation was particularly heavy on the 20th, with falls exceeding 2 inches over a wide area.

'Heavy rainfall occurred in southwest England and South Wales on December 1st, but the period up to the 17th was mainly quiet and at first foggy over England and Wales. Many places in the south recorded no precipitation for a fortnight or more, an unusual occurrence in December.

'Weather became unsettled everywhere from the 18th and some large totals were recorded in the Lake District and in northern Scotland.'

J. Grindley

1972 A relatively dry year. The early months followed the pattern of recent years, a rather mild, damp winter dragging on into spring, with quite chilly weather prevailing till early June. Midsummer brought a series of heavy rainstorms which gave rise to considerable anxiety about harvest prospects. An improvement set in at the end of July, and harvest was one of the easiest and sunniest within memory. The dry spell was prolonged into autumn, though with much cloud and a little light rain. But November and December brought heavy rainfall.

The prolongation of winter into spring resulted in much late sowing, and the weather continued so chilly that some crops never recovered from the bad start. This applied particularly in the western counties, where cereal yields were subsequently down to less than a ton an acre in some places. The chilly dampness of the early and middle months encouraged, too, the growth of fungus diseases. Moulds and blights abounded, though, curiously enough, mildew was not a problem on most cereal crops. Instead, a heavy toll was taken by yellow rust on spring wheat and by rhynchosporium on spring barley, reducing some crops in western districts to yields of approximately half the 1971 levels.

The heavy storms of midsummer spoiled many crops of hay, though the later cuts, especially of meadow hay, proved excellent. Many fields of cereals were flattened by the same storms, but the fine harvesting weather enabled most of the grain to be harvested without much deterioration.

The harvest drought came too late to do much damage to kale and roots, which simply rested for a few weeks and then started to grow again. Grazing grass also produced an excellent autumn crop, and good feeding straw was in abundant supply. Farmers went into winter well supplied with stocks of food.

Most species of aphids were extraordinarily abundant throughout the summer, affecting vegetables most severely, but there was little trouble with potato blight. The cool, dismal weather at blossom time adversely affected fruit trees, which in general produced a poor harvest.

The total cereal acreage, at 9,386,000, was a little below the high level of 1971, and the total yield was about 15,250,000 tons. The barley and oat harvest was heavier than in recent years, but the wheat harvest slightly lower. Prices for cereals were high, reflecting world shortages. The potato acreage was 21,000 below the target of 605,000 acres. The sugar beet acreage remained steady, but there was a decline in yields.

Total numbers of cattle (including calves) continued to increase (to 13,483,000 compared to 12,804,000 in 1971), the main expansion occurring in the beef herd. The dairy herd not only increased by about 100,000 but also gave an improved performance, milk yields reaching record levels. Sheep began to show an upward trend, but the numbers of pigs fell a little. Poultry figures remained fairly constant.

Land prices rose a little (from around £200 an acre to £234, with vacant possession, though with some much higher figures in favoured districts). The number of farm units continued to decline.

LAST YEAR WAS SCOTLAND'S DRIEST SINCE 1955

Reproduced by kind permission of The Times

'Rainfall last year was less than average over England, Scotland and Northern Ireland and slightly greater than average over Wales. It was the driest year over Scotland generally since 1955.

'The first half of the year was generally wet over each country, particularly England and Northern Ireland, where each of the months January to June had greater than average rainfall, but the main feature of the distribution was the very dry late summer and autumn. The period July to October was the driest since 1947 over England and Wales, the driest since records began in 1900 over Northern Ireland and the driest over Scotland, by a wide margin, since records began in 1869.

'A notable feature since 1968 has been an apparent increase in the incidence of heavy rainfall (25 mm or more) covering very wide areas. Three such events occurred in 1968, each unique in different aspects with regard to the area covered by specified amounts. Similar events occurred in 1969 and 1971 and 1972 provided yet a further example on September 8th, when the general depth of rainfall over England and Wales exceeded 25 mm.

'The area of heaviest fall extended from south-west England across Wales and the Midlands to the Humber and Wash. The approximate area covered by falls of 25 mm or more is 78,000 square km. The areal extent, if confirmed, would make this the third largest daily rainfall event for falls of 25 mm or more. Areal coverage for larger amounts was relatively insignificant, the highest daily total being 72 mm.

'Among the largest and smallest rainfall totals recorded

were about 5,100 mm at Styhead, in the Lake District, and 338 mm at Geanies House, in Ross.

'The first few days of January were cold, with rain or drizzle and even a little snow to the east of London on the 2nd. Weather remained unsettled throughout the month and there were no extended dry spells. Heavy snowfall occurred on the 17th in the Scottish Highlands and further snow fell in western and northern districts from 18th to 21st.

'The last few days of January and the first few of February were extremely cold and heavy snowfall (depth up to 8 inches) occurred in Kent on January 29th. February 1st was also a snowy day over much of the country. The rest of February was mainly snow free but unsettled. Precipitation was frequent and heavy up to the 18th, the 15th being very wet in Glamorgan with 93 mm at Treherbert. The period 19th–27th was drier generally but heavy rainfall occurred in the extreme southwest from 23rd to 29th; and the 27th was a very wet day in Down where 85 mm was recorded at Foffany reservoir.

'In March, most of the precipitation occurred in the first and last weeks, the first week being particularly stormy. Many places experienced up to a fortnight with little or no rainfall in mid-month. Precipitation was wintry in character in Britain about 3rd–5th and 26th–28th, with blizzards in Scotland in the latter period.

'The wet and unsettled weather of the last week of March continued for the first half of April with some thunder about 10th–11th. Over large areas of the country, including Scotland and Northern Ireland, a fortnight of almost unbroken dry weather occurred from 14th to 27th, but the last few days were wet generally.

'May was generally unsettled and windy. The period of heaviest rainfall over England and Wales occurred from 5th–12th, with widespread but not intense thunderstorms about 10th–11th. A generally dry period extended from 13th to 19th over England and Wales and up to 22nd over Scotland and Northern Ireland. The period 24th–30th was generally wet over Scotland, and one of the wettest days of the year occurred there on 25th; falls exceeded 75 mm at widespread places, the greatest daily total being 119 mm at Tyndrum, Perthshire.

'June was unsettled and very cool almost throughout. The period up to the 12th was the wetter part of the month over

England and Wales but the 17th to 27th was much wetter over Scotland. Rainfall in eastern England was mainly slight from 13th but more substantial amounts were recorded in the hilly west. The heaviest point rainfall of the year occurred on 17th when 185 mm was reported at Honister Pass, Cumberland.

'Cool, unsettled weather continued for the first eight days of July but the period 9th to 29th was mainly dry and warm over England, Wales and Northern Ireland; the start of the dry period was delayed until 13th over Scotland. The 3rd was a very wet day in North Wales, 105 mm being recorded at Blaenau Ffestiniog, and Honister Pass again recorded an outstandingly heavy fall with 110 mm on 4th.

'Notable exceptions to the dry weather in mid-month occurred with an outbreak of locally very intense thunderstorms in England from 18th to 24th. In this period more than 40 falls attaining the category 'noteworthy' (defined as a fall of given amount and duration expected to occur once in 10 years or less frequently) were recorded and of these six attained the category 'very rare' (return period once in 160 years or less frequently).

'The outstanding fall was at Exeter airport on 18th where 89 mm was recorded in 132 minutes. East of Exeter rainfall exceeded 80 mm over an area of about 100 sq km. Further heavy thunderstorms were reported in the Midlands on 23rd–24th. An even heavier fall, but of longer duration, occurred at Costessey on 31st (in the early hours of August 1st) when 140 mm was recorded in $3\frac{3}{4}$ hours. The epicentre of the storm lay a little to the west of Norwich, and eastern parts of the city escaped with small amounts of rainfall.

'Further heavy thunderstorms occurred on August 1st, notably in Kent, Worcestershire and Derbyshire, and weather continued unsettled and rather wet up to the 7th. The notable feature of the month was the long dry period which extended from August 9th to September 6th over England, Wales and southern Scotland (the north of Scotland was rather more unsettled); most places experienced a little precipitation during the period but locally up to 30 days without measurable rainfall were reported.

'The dry weather was broken on September 7th and the wettest day of the year occurred on 8th over England and Wales when a general depth of 28 mm was estimated; rainfall in Scotland and Northern Ireland was comparatively trivial on that day. The long spell of dry weather was resumed from about September 14th and continued almost unbroken over the whole country up to October 7th. A notable break occurred in south-east England and East Anglia on 17th–18th when values exceeded 25 mm at places, the highest total being 65 mm at Southwold, Suffolk.

'Most of October's rain fell in the periods 8th to 10th and from 26th, that is to say a further fortnight of rainless or very dry weather occurred in mid-month over much of England, Wales, Northern Ireland and southern Scotland. In all, from August 9th to October 25th many places experienced 11 weeks with only small amounts of rainfall (less than 50 mm) although nowhere was completely rainless in that period.

'In parts of East Anglia rainfall was nil for the whole of October but at most places the 26th marked the end of the long dry period, November in general being a wet and stormy month. Most of the precipitation occurred in the period 4th–20th, with the 9th and 12th in particular being very wet. Honister Pass again reported a very heavy daily fall of 172 mm on 9th; 116 mm was recorded at Thirlmere and at many places from Caernarvonshire to Westmorland more than 75 mm occurred on the same day.

'The 18th was a wet day in Northern Ireland; Foffany reservoir recorded 72 mm. The 28th to 30th were all wet days in hilly western districts, the highest total being an outstanding fall of 158 mm at Honister Pass on 29th (with a further 58 mm there on 30th). Some mainly light snowfall and occasional thunderstorms were reported in the period 10th–18th.

'December weather had two distinct phases. The period up to the 12th was stormy and wet with many places recording more than the average for the month by that day. An unusually large number of thunderstorms were reported in England and Wales from 4th to 8th and the period 4th–10th was snowy in Scotland. Very heavy rainfall was reported in the west on 1st, 2nd, 5th and 11th, a fall of 99 mm occurring on 11th at Honister Pass. The period from the 14th was mainly dry, quiet and rather foggy.

'Most places recorded small amounts of rainfall but some parts of the north-east and East Anglia experienced two or even three weeks without measurable precipitation. Moderately heavy rainfall was experienced in south-west

England on 26th–28th and weather in Scotland and Northern Ireland was generally unsettled from 23rd.'

J. GRINDLEY

1973 1973 was a year difficult to fault, as far as weather and the performance of crops and livestock were concerned. Farming problems were nearly all man-made.

The early months brought no unduly hard weather, nor were there late frosts and icy winds in spring, to damage fruit blossom and check growth. Enough rain fell in summer to keep crops growing. These summer storms certainly spoiled some hay, but most farmers managed to make ample for their needs. The same storms gave cereal crops a battering, but sunny weather in August and September corrected the damage, and yields suffered little. Harvest was, in fact, one of the easiest in recent years, and much grain went into store with little or no drying. Barley crops gave fair yields, but wheat was well above average. In spite of the mild winter, little disease appeared.

In the open autumn excellent progress was made with ploughing and autumn sowing. Winter corn germinated evenly and at the end of the year was looking well. So farmers finished 1973 with the seasonal work of the farm nicely up to schedule and with ample stocks of home-grown food for their livestock, which were looking very fit. There were no major outbreaks of animal disease during the year, though the irritating swine vesicular disease cropped up sporadically.

The total cereal harvest was nearly 9,300,000 acres, a little below the levels of the previous two years. Total yields were just over 15,000,000 tons. Prices, however, rose to record levels, owing to inflation and world shortages of grain. The potato acreage fell 24,000 short of the target of 580,000 acres, but the average yields were about 0 75 tons higher than in the previous year. Sugar beet acreage and yields were about normal.

The national cattle herd, both in dairy and beef, continued to increase, the total of adult cattle and calves being 14,445,000. The beef breeding herd expanded by 14%, the dairy sector by 3%. With the prices of imported feeding stuffs rising dramatically in autumn a considerable cutback in the use of concentrates occurred, with consequent fall in milk yields. The numbers of pigs, sheep and poultry all increased slightly during the year, and prices remained firm, though costs, particularly of feeding stuffs for pigs and poultry, rose rapidly.

The trend towards larger and fewer farming units continued, the estimated number of full-time farms being 178,000, compared to 196,000 in 1968. The average price of land sold more than doubled, soaring from about £200 (the average for 1969 to 1971) to more than £500 by the end of September.

Among minor crops, oilseed rape increased its acreage considerably, from 17,000 acres in 1972 to 34,000 in 1973. The acreage of grass cropped for herbage seed also increased.

In this year the United Kingdom entered the European Economic Community.

LEAST RAINFALL IN ENGLAND AND WALES SINCE 1964

Reproduced by kind permission of The Times

'Rainfall last year was less than average over the United Kingdom generally and there was only a 6 per cent spread between the highest and lowest percentage values of each country. Over England, Wales and Northern Ireland the rainfall map was featureless, but over Scotland, where percentage annual values were from more than 130 to less than 60, it was noteworthy that the general value was so near that of the other countries.

'Over Scotland and Northern Ireland nine months of the year were drier than average, but over England and Wales the distribution was characterized by the rather wet summer and the dry winter months, especially those early in the year. Over England and Wales it was the driest year since 1964 and there had been only three drier years this century.

'One of the most important features was the accumulated deficiency of rainfall over England and Wales, more especially over eastern England, and also over eastern Scotland. The deficiency of rainfall over England and Wales compared with average began in July, 1972; by the end of March, 1973, the nine-month total was the lowest since 1750.

'In the winter half-year storage is normally replenished, but the rainfall from October, 1972 to March, 1973, was barely sufficient to raise ground-water levels, especially in eastern districts. Apart from August, the summer months were all wetter than average but not sufficiently so to raise the water levels. The last three months of the year were again dry, and there remained an accumulated deficiency of rainfall over much of England, Wales and east Scotland.

'Among the largest and smallest rainfall totals recorded were about 4,000 mm at Ben Moore, in Sutherland, and 329 mm near Biggleswade, in Bedfordshire. It is probable that about 4,500 mm fell in Glen Quoich, Inverness-shire.

'January began with a few days of unsettled weather but that was followed by more than a week of mainly dry weather. For the rest of the month unsettled weather returned to the north and west of the country and particularly heavy rain occurred in Northern Ireland and western districts on 19th and 20th, and snow blocked roads in parts of Scotland, northern England and Wales; daily rainfall exceeded 100 mm in co Down on 19th.

'The first and third weeks of February were relatively dry over England, Wales and eastern Scotland. The other weeks were unsettled everywhere, with wintry showers sometimes prolonged and thundery. Between 21st and 23rd parts of Scotland experienced the worst blizzards for three years. There was rain in most places on the first day of March and that was followed by showery weather until 6th; the showers became wintry on 5th and 6th.

'It was dry over most of the country from 7th until rain spread from the west on 23rd and 24th. Thereafter showery weather prevailed but the 26th and 28th were dry in most parts. March was the third successive winter month with less than normal rainfall, and concluded the driest winter half-year, October to March, over England and Wales since 1809–10.

'April 1st was a wet day everywhere with strong to gale force northerly wind, especially over England and Wales, where it heralded the first month of 1973 with above average rainfall. The weather remained unsettled until 8th, then gradually became drier. Easter weekend was wet over England and Wales with thunderstorms in places but it was drier over Scotland and Northern Ireland. It became drier again immediately after Easter Monday but unsettled weather returned at the end of the month and it finished wet as it had begun.

'May Day was dry apart from northern Scotland but the

month generally was wet, the only dry spells being 11th and 12th in England and Wales, 14th to 18th over the whole country and 24th to 27th over England and Wales. On May 3rd widespread thunderstorms affected southern England, causing floods in many parts; Sussex experienced another flood on 6th. Further thunderstorms occurred on 21st and 28th, mainly over the Midlands and again with reports of flooding.

'June, over England and Wales, was a month of long dry periods, broken by short thundery spells. The rain associated with those thunderstorms was often heavy and over much of England more than the average amount of rain for the month fell on those few wet days. There were some notable occurrences. On 19th general rain over England and Wales as a whole exceeded 23 mm in 24 hours at Chessington, Surrey. On 27th thundery rain was rather more localized and near Didcot, in Berkshire, 66 mm was recorded in six hours. Over Scotland rainfall was more evenly distributed during the month. In Northern Ireland it was the driest June since 1949, although 1697 was equally dry.

'The pattern of weather was basically the same in July as in June. Over Scotland and Northern Ireland rainfall was less than normal but there was only one prolonged dry period from 23rd to 28th, over Scotland and from 25th to the end of the month in Northern Ireland. Over England and Wales there were long dry spells at the beginning and end of the month. Apart from thundery outbreaks on 6th it was dry in most places until 12th and again from 23rd to 31st.

'On 6th rain was very heavy in parts of east and south-east England; more than 100 mm occurred in parts of Surrey. In the unsettled spell from 13th to 22nd thunderstorms were reported from many places with local flooding, but the notable wet day of the month was 15th, when violent thunderstorms occurred in the area of the southern Pennines; daily values exceeded 50 mm in many places and 100 mm near Sheffield; at Rivelin, west of Sheffield, 155 mm fell in the two days, 15th and 16th, more than twice the average for July.

'August was generally unsettled until 7th, and in the north until 10th. Heavy rain was widespread on August 5th, which was the wettest day of the year over England and Wales, with more than 25 mm generally; several daily falls in excess of 100 mm occurred in Wales. From 8th, apart from isolated thunderstorms, it was predominantly dry until the end of the month, especially over south-east England, where there was almost no rain in some places. The end of the month was somewhat wetter in the north but amounts were not large until 31st.

'The weather of September was in two phases. The first half of the month continued the dry spell of late August. On 14th thundery rain broke out in south-west England and spread to most of the country on 15th. The rest of the month was unsettled. Rainfall was very heavy in thunderstorms over south-east England on 20th, particularly in east Kent, where some daily values exceeded 150 mm.

'The first five days of October were dry in most places. Scattered heavy thunderstorms broke out over England on 6th and a fall of 40 mm in 35 minutes was recorded near Maidenhead, Berkshire. The weather then became unsettled, but not without short dry spells; heavy rain affected Cornwall on 12th, with more than 90 mm in some places; snow fell in Scotland between 16th and 18th. From 23rd it was dry until the end of the month over England, Wales and eastern Scotland.

'Unsettled weather predominated throughout November in Scotland and Northern Ireland, although the second half of the month was considerably drier in eastern Scotland. It was the driest November over England and Wales since 1956 and most of the rain, or snow late in the month, fell on isolated days, 4th, 9th, 14th and 28th and in south-west England on 30th. Some daily values exceeded 100 mm in north-west Wales on 9th and in Inverness-shire on 17th.

'Over Scotland December was a wet month. Apart from 1st there was no other day in the month without some precipitation, which was mainly snow from 6th to 18th. In England, Wales and Northern Ireland there were few completely dry days; nevertheless there were few really wet ones. On 7th there was a lot of rain, with snow in the north; on 15th heavy snow fell in north-west England and on 19th rain was heavy in western districts.'

JANE AYRES

1974 Following a familiar pattern, 1974 began with a mild winter and a dull, late spring, though without any savage May frosts to damage fruit blossom. The usual May drought was more prolonged and absolute than normal, and late-sown crops germinated badly. After the first flush, spring grass recovered slowly from grazing, and silage and early hay cuts were exceptionally light, though the quality was high.

Late June and July brought their quota of storms, resulting in some damage to crops. August gave a patchwork pattern of rain and sunshine, though with enough of the latter for a good proportion of the corn harvest to be gathered in excellent condition. A few lucky and/or energetic farmers in early districts managed to finish harvest in this benign period. Fortunately the wheat harvest was comparatively early, owing to heavy sowings in the previous mild autumn and the early growth made during the mild winter.

The fine weather broke early in September and never really recovered. A series of violent gales, with much rain, during September was almost unprecedented. The part of the harvest that remained unfinished dragged on into November, and the quality of the grain rapidly deteriorated. On the credit side, the rains helped autumn grass and fodder crops, which brought relief to anxious livestock farmers.

Livestock farmers in general had a bad year, caught between the millstones of falling prices for their products and the soaring costs of feedstuffs. The selling off of stock at times reached panic proportions, as when, during the summer drought, healthy calves went for as little as £1 apiece. In December calves were being sent to the abattoirs at four times the normal rate, and breeding stock was being slaughtered wholesale.

On the other hand, prices were so low that farmers tried to keep as many animals as they could feed, from their own resources, throughout the winter. Most farms, gambling on a reasonably mild winter, entered the winter stocked to capacity, in spite of the drastic culling. But none was willing to risk having to buy much foodstuff, with hay making £2 to £3 a bale.

All this happened against a background of raging inflation.

The cereal harvest at around 16,000,000 tons was a record, and prices remained high. The potato yields were, in spite of the summer drought and difficult harvesting, better than usual, but those of sugar beet were the lowest for many years. Despite the slaughterings, the national cattle breeding herd increased by 3%, a slight fall in the dairy herd being more than compensated for by a rise in the numbers of beef animals. The numbers of sheep also increased, but the pig breeding herd continued to decline, showing a fall of 12% over the 1973 total.

In general, net farm incomes are calculated to have fallen by 17% in terms of money in the year 1973–74; in real terms, allowing for inflation, the fall was 30%.

LAST AUTUMN WAS THE FIFTH WETTEST ON RECORD FOR THIS CENTURY

Reproduced by kind permission of The Times

'General rainfall last year was above average over Great Britain and almost exactly average in Northern Ireland. The year began mild and wet. The rest of the year, except December, was cold with a notably dry spring and wet autumn. Over England and Wales it was the driest spring (March, April and May) since 1956. The combined April–May period was the driest over England and Wales since 1896, over Scotland since 1946 and over Northern Ireland since 1957. The autumn (September, October and November) was the wettest over England and Wales since 1960 and the fifth wettest this century; over Scotland and Northern Ireland it was wetter in 1970 although rainfall was not a great deal above average over Northern Ireland. By the end of December it was again wet and windy but mild, and the old year went out as it had begun.

'Among the largest and smallest rainfall totals recorded were 4717.9 mm at Llydaw Mine, in Snowdonia and 502.6 mm at Dunbar, in East Lothian.

'During January there were long spells of unsettled, rainy weather with strong wind, but for most of the month it was very mild. It was the wettest January generally over Wales and Northern Ireland since 1948. Drier spells occurred in the first few days of the month over England and Wales and more generally between the 17th and 21st, although north and west Scotland remained wet. On the 17th continuous heavy rain produced flooding in Scotland, mainly in an area to the north of Glasgow, and the daily amount of 238.4 mm, recorded near Loch Sloy is the highest daily rainfall ever recorded in Scotland.

'The unsettled wet weather continued into the first half of February. In general, except for north Scotland, more than three quarters of the precipitation fell in the first 15 days of the month. Snow and sleet occurred widely on the 5th and 6th and continued until the 8th in the north. The second half of the month was comparatively dry, apart from the 28th, when widespread rain or snow occurred. It was the wettest combined January and February period over Scotland since 1962, over England and Wales since 1951 and over Northern Ireland since 1937.

'March was the first month of the year with below average rainfall but it was not as dry as March, 1973. Most of the rain (or snow) fell between the 9th and 19th in England and Wales and the 12th and 21st in Scotland and Northern Ireland. After that it became very dry with only a little rain at times here and there; the last two days of the month were completely dry throughout the United Kingdom.

'With one or two interruptions the dry weather of the last 10 days of March continued throughout April. Over the end of March and beginning of April some places experienced three weeks without measurable rain. It was the driest April in Scotland since records began in 1869; in Northern Ireland it was the driest since 1954 and in England and Wales since 1957 and in those countries there have only been three drier Aprils in this century.

'Although the total rainfall for May was below average, except in Northern Ireland, there were few completely dry days in any of the constituent countries of the United Kingdom. Heavy rain occurred over south-west England on the 2nd, thunderstorms were widespread over England on the 23rd and heavy rain was recorded in the north of Scotland over the 27th–28th.

'In June the south of England was very wet, with more than twice the normal rainfall in parts of the south-east; in the north of England, Wales, Scotland and Northern Ireland and especially the last named, rainfall was less than average. Outbreaks of thundery rain interrupted dry weather from the 11th to 24th over the whole of the United Kingdom. The excessive June rainfall in southern England was largely due to heavy rain, which followed this dry spell, while in northern England, Scotland and Northern Ireland it remained dry almost to the end of the month.

'In Wales July was wet, the wettest since 1958, but elsewhere it was drier than normal. Nevertheless, in Northern Ireland it was, apart from 1970 with slightly more rainfall, the wettest July since 1960. In August less than the normal rainfall occurred in Northern Ireland, Wales, Scotland and Northern England, but the general rainfall in England and Wales combined was greater than average, largely due to rainfall much above average in the south and east of England. September was a very wet month throughout the United Kingdom, with nearly twice the normal rainfall in England and Wales generally, while more than three times the average rainfall was recorded in parts of southern England. It was the wettest September in Northern Ireland since 1962 and in England and Wales since 1968. The first week was very wet, especially in southern England and south Wales.

'In October there was less than average rainfall in Northern Ireland, Scotland and Wales but well above average in England. Over England and Wales combined rainfall exceeded the average generally; it was the wettest October since 1967 and the first since then with more than average rainfall. November was wetter than normal throughout the United Kingdom, but only slightly so over Northern Ireland, where it was the driest November since 1968. Again, as in the preceding month, there were no completely dry days.

'The rainfall of December was less than the average over England, average over Wales and above average over Scotland and Northern Ireland, especially the former; in the last two countries it was the wettest December since 1966, and in Scotland there have only been two wetter Decembers this century. The first nine days of the month were mainly dry over England and Wales. The 10th was a wet day everywhere, with snow in most parts of Great Britain except the south-west; thunder was reported from south-east England. The rest of the month was unsettled, Christmas and Boxing Days being particularly wet but mild without the 'traditional' snow except in Shetland.'

JANE AYRES

1975 The outstanding event of this year was the subtropical summer, when throughout most of June, July and August the temperature stayed in the 80°s and 90°s F and the sun beamed down from a cloudless sky.

The first few months followed the pattern made familiar in previous years, namely, mild, damp and cloudy. Unfortunately it persisted late, without even the usual break offered by a spell of cold, drying winds in March. Spring sowing was, in consequence, exceptionally late, expecially on the heavier soils. Many fields were not planted till late April and even early May.

The change to drought and heat eventually occurred suddenly. The summer was not without brief stormy intervals, but when the heat and sunshine returned the soil soon dried out. Throughout almost the entire summer, therefore, crops and all vegetation were contending with a chronic shortage of moisture.

Harvest started and finished early, leaving some farmers satisfied and some talking of disaster. The demarcation line lay between those on light soils and those on heavy; between those who had sown most of their acreage to autumn corn and those who had relied on spring crops. Fortunately, the previous autumn had been unusually favourable to cultivations, with the result that more winter wheat and winter barley than usual went in. These crops produced good yields. On the other hand, some of the late-sown spring barley was hardly worth harvesting. Yields of 10 to 15 cwt per acre were reported, and some barley crops were either ploughed in or cut for silage.

The grass cut for hay and silage was very light, but fortunately the weather held fair for haymaking and therefore the quality was excellent. There was no recovery after the early cuts, however, and throughout the summer most pastures looked as bare and brown as deserts, though cattle and sheep continued to thrive on the short, close-nibbled herbage.

The September rains, though not excessive, produced an extravagant response. Cows were soon knee-deep in grass and clover. Crops of kale and maize, which had been marking time for months, suddenly achieved phenomenal growth. Farmers took advantage of the early harvest to sow catch-crops in stubbles, and fodder turnips and rape in particular produced tons of food per acre in the course of a few weeks. A well-organised straw lift brought thousands of tons of good quality straw from the arable east to the badly-hit west, so, all in all, few inroads were made into conserved fodder during autumn. Farmers went into winter more comfortably placed than they had expected. The panic selling of livestock, a feature of the autumn of 1974, was not repeated, and sales and slaughterings eased off to normal.

The cereal harvest of 13,800,000 tonnes was well below the record of 16,400,000 tonnes in 1974. The decrease in wheat acreage, owing to difficult sowing conditions in the autumn of 1974, was not entirely compensated for by spring barley. Yields of potatoes and sugar beet were much below normal. With beef cattle prices high and the outlook for milk somewhat discouraging, autumn marketing of fat cattle reached an all-time record. The numbers of sheep and pigs showed a further decline.

The Government issued a White Paper, 'Food from our own resources', setting out guidelines for objectives and policies to the 1980s.

1975 WAS FIFTH DRIEST YEAR OF THE CENTURY

Reproduced by kind permission of The Times

'Rainfall last year was well below average over the United Kingdom generally, but 1973 was drier. It was the fifth driest year of the century over England and Wales and the second driest over Northern Ireland. Drier years were 1971, 1972 and 1973 over Scotland.

'The monthly rainfall distribution showed a wet January and a wet September over the country generally and the early spring was also wet over England and Wales. A notable feature of the distribution was the long, dry (and often hot) summer over the United Kingdom. The dryness of the last three months of the year, following on the dry summer, may cause concern for the water supply in some parts of the country.

'It was the wettest January since 1928 over Scotland and the three months November, 1974, to January, 1975, were the wettest such months over that country since comparable records began in 1869. There has been no drier May–August over England and Wales since the summer of 1959; other drier periods May–August were those of 1911, 1940 and 1949. It was easily the driest May–August over Northern Ireland since records began in 1910 and the driest since 1955 over Scotland. Perhaps most remarkable of all was the dryness of

October–December; there has been no drier such period since 1879 over England and Wales, since 1937 over Scotland and since 1922 over Northern Ireland.

'Despite the dryness of the summer there were few extended spells without measurable rainfall; at most stations the longest spells were between two and three weeks, mainly ending about July 7 over much of England and Wales. Exceptionally, up to five weeks without rainfall occurred in Derbyshire, Dorset, north Devon and west Wales.

'Among the largest and smallest rainfall totals recorded were 4028 mm at Delta, Snowdonia and 357 mm at Mexborough, Yorkshire.

'Predominantly westerly weather prevailed throughout January, although under the influence of high pressure to the south of the country the first 11 days were mainly dry over England and Wales. Scotland experienced wet weather from the start. Flooding followed heavy rainfall about January 19–21 in many parts of England. Although it was a mild month generally snowfalls were fairly widespread in Scotland between January 16 and 28.

'Most of the rainfall of February occurred in the middle 10 days. The 1st to 9th was a predominantly dry period throughout the United Kingdom as was the 19th to 28th over England and Wales. Amounts of rainfall were not unduly large although the 16th was a generally wet day over Scotland. It was the driest February since 1965 over the country generally.

'The first week of March was unsettled and wet over the whole country and the unsettled weather continued until the 13th over southern Britain. The period 8th to 20th was mainly dry over Scotland and Northern Ireland as was 14th to 25th over the south, but the 21st was a generally wet day over the whole United Kingdom. From the 26th a northerly airstream brought snow showers, some of which were heavy, to most parts of the country and Easter was wintry.

'The first three weeks of April were unsettled over most of the country, and precipitation was wintry, particularly on the eastern side of Britain for the first 10 days. Milder weather spread over the country on the 11th and it became much drier over southern Britain from the 19th and over the North and Northern Ireland from 22nd. Cooler, showery weather returned to most areas in the last few days of the month.

'After a rather wet start in the North, May was dry nearly everywhere in the country. The 7th was wet, particularly in the Midlands and East Anglia, and the period up to the 14th was unsettled. Flooding followed heavy rain in south-east England on the 16th. The rain was associated with a depression over France; more than 50 mm was recorded in Sussex and Kent. All areas had a long mainly dry spell from the 18th until the end of the month.

'June was notable for the unusually late snowfalls which occurred on the 1st and 2nd over the country including parts of the South; the snowfall was the first in June since 1888 in the London area. Frontal rain affected northern England and Scotland on the 3rd–4th but the 3rd to 14th was mainly dry over England. The period 15th to 17th was showery and thundery but the rest of the month from the 18th was dry nearly everywhere apart from heavy rain in the London area on the 23rd. No rainfall at all was measured throughout the month in parts of Dorset and Hampshire.

'The dry weather of the second half of June continued until July 6th over most areas except the extreme north-west. Isolated heavy thunderstorms ocurred on the 4th and the dry spell was broken generally on the 7th, when a depression moving north from Biscay brought thunderstorms to many parts of England; Scotland and Northern Ireland remained dry until the 9th. Showers or longer periods of rain affected most areas until the 24th but amounts were not generally large except in local thunderstorms between the 9th and the 18th. Remarkably heavy orographic rainfall occurred in the Lake District and Pennines on 21st to 23rd; at Honister House, Cumbria, 220 mm was recorded on the two days 21st to 22nd. Weather over England and Wales was mainly dry and sunny in the last week but more unsettled over Scotland. At Medstead, Hampshire, 96 mm fell in a thunderstorm on the 31st.

'The dry weather of late July continued generally up to August 13th although in the intense heat local thunderstorms occurred and were often heavy, particularly on the 4th and more generally on the 8th, when 97 mm were recorded at Goonhavern in Cornwall. Thunderstorms were again widespread on the 14th and on that day one of the most remarkable events in British rainfall history occurred. At Hampstead 171 mm fell in about 2½ hours in an intense, thunderstorm (the area of the heavy rainfall was remarkably restricted). That is by far the heaviest fall in a day in the London area and the magnitude of the fall may be judged by the close approach to the calculated maximum possible fall (190 mm) for a duration of two hours at any one place in the London area.

'After the 15th weather became cooler, with occasional rain or showers in many districts until the 23rd. Heavy rainfall occurred in south-west England on the 15th to the 18th, and the 19th was a wet day over the country generally. The last week was mainly dry but heavy rain fell in northern England and southern Scotland on the 29th, 108 mm being recorded at Sweethopes Loughs, Northumberland. At Mileham, Norfolk, rainfall for the month was only 1 mm.

'The long dry summer came to an end in September, although the first week continued mainly dry over England and Wales. The rest of the month was generally unsettled, with many days of heavy rainfall and occasional clear dry days. The 13th was a very wet day in southern Britain as a depression moved up the English Channel; rainfall exceeded 50 mm over 6,265 sq km. The 16th and 17th were very wet in southern and eastern Scotland and Northern Ireland. The 22nd was wet in northern Scotland and 24th to 27th was very wet over the country generally. The wet weather of September continued up to October 2 over England and Wales and up to the 4th over Scotland and Northern Ireland. The rest of the month was mainly dry, although rainfall occurred about the 8th to 10th (heavy in south-west Scotland, Wales and south-west England on the 9th) 15th to 17th (mainly over England).

'The 22nd and 23rd were very wet days over western Scotland and Northern Ireland, and north-west Scotland remained wet until the end of the month. Heavy rainfall occurred in south-west England, Wales and south-west Scotland on the 31st.

'November was mainly unsettled particularly over Scotland. The wettest day of the month over England and Wales generally was the 2nd and rainfall was mainly light over those counties from the 3rd to the 14th and over Scotland and Northern Ireland from the 6th to the 14th. The 15th was an extremely wet day over the whole country, the cause being a rapidly deepening depression which moved eastwards across Scotland. Weather continued unsettled for the rest of the month, apart from brief drier spells over much of England and Wales from the 20th to the 23rd and 29th to 30th. The 26th to 28th was a very wet period over Scotland, much of the precipitation falling as snow.

'The 1st was by far the wettest day in December, falls exceeding 25 mm over a wide area and over 50 mm in north-west England. The precipitation occurred as snow in midland and eastern England and southern Scotland. The period from the 3rd to the 22nd was mainly dry over England, Wales, Northern Ireland and southern Scotland, with occasional rain or snow; northern Scotland was much more unsettled and wetter in that period, with snow between the 11th and 17th. The 23rd and 24th were generally wet days with snow in northern Scotland. The 25th to 29th was again mainly dry over England, Wales and southern Scotland but the 29th was a very wet day over north-west Scotland. During the last two days of the year rain swept south over the whole country, and thundery snow showers occurred in northern Scotland.'

J. GRINDLEY

1976 The year started in pedestrian fashion, with a mild winter and late spring, much on the pattern of the preceding years. Then came an unprecedented drought, with temperatures in the high 90° F spectrum day after day. It was a glorious summer, the sort of golden season that lives in the memory for a lifetime.

Suddenly, it was over. In early September, with hardly an interlude, the weather switched to monsoon conditions, causing floods before the water authorities had got round to cancelling arrangements for combating the drought. Excessive rains confined tractors and other farm machinery to their sheds for much of the autumn, while the sodden fields went unploughed and potatoes rotted in the ground. In December frost set in while the countryside was still saturated. Fog, ice and treacherous roads produced hazards not experienced in December for many years.

Harvest yields were disappointing. Some spring-sown crops had no rain at all from seed-time to harvest. Grass was luxuriant in spring and early summer, and early cuts produced hay and silage of excellent quality. The haymaking season was ideal. Thereafter the pastures dried up till they resembled threadbare doormats, and farmers began to doubt whether they would ever recover. Kales, swedes and other crops for autumn fodder failed, and many livestock farmers had to start using their winter supplies of hay and silage in July and August. But when the autumn rains began to fall on the superheated soil vegetation soon started to grow luxuriantly again.

The total area of cereals harvested in the United Kingdom amounted to 9,090,000 acres, against 9,020,000 acres in the previous year; but production, at 13,450,000 tonnes, was about 500,000 tonnes less, owing to the drought. The favourable sowing conditions in the autumn of 1975 caused an increase in the wheat acreage and a corresponding fall in that of barley. Although the potato acreage exceeded the target of 546,000 by some 5,000, the yield was probably less than two-thirds of normal, and prices rose to unprecedented levels. Oilseed rape continued its rise in popularity and occupied the record total of 118,000 acres.

The national dairy herd remained at its 1975 level but the beef herd declined by 7%. Fat cattle were scarcer than usual, owing to the fall in the breeding herd in 1974. Milk yields per cow fell, owing to the drought. Sheep numbers were virtually unchanged, but pigs were recovering from the depression of the earlier years of the decade, the breeding herd expanding by 9%.

The estimated average wheat yield was 1.54 tonnes per acre; barley, 1.42 tonnes per acre; oats, 1.36 tonnes; potatoes, 8.2 tonnes. The average milk yield was 925 gallons per cow.

In general arable farmers did worst in 1976, though high potato prices provided some compensation. Livestock farmers were helped by higher cattle and sheep prices. Horticulture fared badly, and there was a continued fall in fruit and vegetable acreage.

A feature of the mid-1970s was the use of fodder turnips and rape instead of swedes, mangolds and kale. Some of the larger farms began the practice of the aerial sowing of fodder turnip seed in fields of standing barley in July, with gratifying results.

A CONTRAST OF RAINFALL DEFICIENCY AND EXCESS: AUTHORITIES FACED UNPRECEDENTED CONDITIONS

Reproduced by kind permission of The Times

'Although 1976 as a whole was not a notably dry year (there have been two drier in the past six years over England and Wales, and many drier over Scotland and Northern Ireland), there were some remarkable contrasts in rainfall deficiency and excess.

'The long dry period from January to August over England and Wales must inevitably be linked with the dry eight months from May, 1975, as the combined total for the 16 months, over those regions generally, is the lowest in historical rainfall records, which extend back to 1727.

'The drought beginning in May, 1975, can confidently be assessed as the worst on record for England and Wales for a 16-month period for as far back as quantitative comparisons are available. The total for May, 1975, to August, 1976, was 756 mm; the next driest 16-month period starting in May was that in 1749–50, when the total was 809 mm, and the next driest 16-month period starting in any month was that beginning in July 1949 (779 mm).

'The area of most marked deficiencies over the 16 months extended from Devon along the south coast to Dungeness, over South Wales and into the Midlands as far north as Flamborough Head, but excluding much of East Anglia and Kent. In the area outlined, less than 60 per cent of the average for the 16 months was recorded (60 per cent of average in 12 months is considered a rare event) and along parts of the south coast and locally in the Midlands less than half of average was recorded.

'Records were also broken for other periods of months ending in August; the nine, 12, 15 and 18 months totals ending in August, 1976, were unprecedently low in the 250 years. There has been one drier six-month period, March to August, than that of 1976 (in 1741) and one drier three-month period, June to August (in 1880).

'June to August, 1976, was also notable for general warmth

and for the extreme sustained heat of the last week in June and first week in July. The rain total for the 16 months ending August, 1976, was also the lowest in Northern Ireland since before 1900.

'Over Scotland the longer periods ending August, 1976, though dry, were fairly commonplace, recurring on average about once in five or six years, but the three months' total for June to August was more remarkable; in 108 years of comparable records there has been only one drier period (in 1955). Over Northern Ireland the period June to August, 1976, was easily the driest such period since comparable records began in 1900. Total rainfall over England and Wales for the six calendar years 1971–76 was the lowest since the six years 1854–59.

'The long drought with little winter (1975–76) replenishment of reserves, resulted in water authorities being faced with unprecedented conditions. Water supply difficulties and restrictions in parts of England and Wales ensued. The ending of the drought was as dramatic as its severity. Over an area extending from south-west England across South Wales into East Anglia, it may be said to have ended with the heavy storms at the end of August and its termination was marked decisively over the rest of England and Wales by the heavy rainfall beginning in the second week of September.

'Remarkably, the combined total for September and November, 1976, over England and Wales was the highest for such a period in the 250-year record, and that for Northern Ireland was the highest since before 1900.

'Among the largest and smallest rainfall totals recorded were 3,881 mm at Styhead, in the Lake District, and 370 mm at Southminster, Essex.

'The year began with heavy precipitation, much of it snowy north of the English Midlands, and January 2 was probably the wettest day of the year over Scotland generally. Weather remained generally unsettled over Scotland and Northern Ireland and there was further snow in those countries from the 18th to the 28th.

'The period from the 5th to the 17th was generally dry over England and Wales but cold northerly winds brought snowfall, mainly slight, to those areas from the 23rd to the 26th. The last few days of the month were dry nearly everywhere except in extreme south-west England, where heavy rainfall occurred from the 28th to the 30th.

'The cold spell which began about January 23, continued up to February 5 with widespread, light snow showers over England and Wales. Precipitation occurred on many days but was mainly light. The periods from the 13th to the 19th and the 24th to the 28th were mainly dry but the 22nd to the 24th was wet in Scotland with more than 100 mm recorded in parts of the Highlands on the 24th.

'The first week of March over Scotland and the first 10 days over England and Wales were mainly dry with a few light snow showers. The period from the 11th to the 16th was generally wet, the 12th particularly so in south-west England. The week of the 18th to the 23rd was dry over much of England but the 20th and 21st were exceptionally wet in the west; more than 125 mm was recorded in the Mourne Mountains on the 20th, which was also a wet day in the Isle of Man, and more than 80 mm in Devon and Cornwall on the 21st.

'The month remained unsettled, with wintry precipitation in Scotland, but England and Wales were mainly dry from the 26th. Rainfall occurred mainly in small amounts over England and Wales in the first 12 days of April, but Scotland had more generally unsettled weather. The 13th and 14th were mainly wet but the second half of the month was almost rainless over England, Wales, Northern Ireland and southern Scotland.

'In May small amounts of rain, some of it thundery, fell widely. Weather remained unsettled throughout, particularly in the north and in Northern Ireland, but there were several periods of two or three dry days in the south. June was mainly dry over England and Wales, apart from the third week, when the heaviest rain fell between the 16th and 19th; the 19th was a particularly wet day, especially in south-west England, where 68 mm was recorded in parts of Devon.

'In Scotland the first week was mainly dry but the 9th to the 22nd was wet (particularly the 11th) and the wet weather continued until the end of the month in the north. In many widely scattered places in England and Wales rainfall was negligible (less than 0·5 mm) throughout the month.

'The first half of July was mainly dry in most districts, although heavy rain fell in North Wales on the 4th, Tyrone on the 5th and northern Scotland on the 9th. Heavy local thunderstorms occurred in Northamptonshire on the 12th and rainfall was widespread and thundery in southern Britain on the 15th. Heavy thunderstorms on the 20th were mainly confined to south-east England, where more than 50 mm was recorded in Essex and Kent.

'Most places experienced dry weather from about the 19th, although some rain fell in East Anglia, north-east England and north-west Scotland in the last few days of the month.

'In August occasional thundery rain occurred in some areas in the first few days, but most of the country was dry up to the 27th. In the last few days of the month sporadic outbursts of thundery rain, often very heavy and prolonged, occurred in many parts of England and Wales, the areas most affected lying from south-west England through the southern Midlands to East Anglia.

'On the 19th 88 mm was recorded in Herefordshire and on the 30th 106 mm fell in Cambridgeshire. The longest period of rainless weather in the year occurred from mid-July to near the end of August; many places experienced up to four weeks without rain and in parts of the South-west as many as six weeks were rainless. In August less than 1 mm of rain was reported in many parts of England and in the Scottish Highland region.

'Apart from a little rain on the 1st, the first week of September was mainly dry. From the 8th (7th in Scotland) to the 14th outbreaks of heavy, prolonged rain occurred in all areas; north-east England was most severely affected, with 97 mm on Silpho Moors on the 9th and 145 mm (the heaviest daily fall of the year) in the Cleveland Hills on the 10th, which was generally the wettest day of the year over England and Wales.

'The period from the 15th to the 19th was generally dry, but a slow-moving front brought heavy rain to western Britain on the 20th and from the 22nd rainfall was frequent, often torrential and thundery.

'On the 28th flooding came after a fall of 86 mm in four hours in Glasgow. Daily totals exceeded 50 mm in different counties of the United Kingdom on the 8th to the 11th, the 14th, the 21st, the 24th to the 25th, and the 27th to the 30th. Amounts approached 100 mm in Antrim on the 10th.

'October was an unsettled month almost throughout. In particular, the first week saw the continuation of the very disturbed wet weather of late September. There were a few dry days over much of England in mid-month and towards the end of the month. The 14th was a wet day over the United Kingdom generally, with amounts exceeding 50 mm in counties extending from Down to Northumberland and Devon to Lothian; in Snowdonia 114 mm was recorded. In the Mourne Mountains 120 mm fell on the 17th.

'At first in November sunny periods alternated with longer periods of rain, which was often thundery. In general, amounts were not as great as in the preceding two months but rainfall was very heavy in south-east England from the 7th to the 9th and again in the south on the 11th. Weather was mainly dry from the 18th to the 25th but with the resumption of disturbed westerly type weather the last few days were generally wet, particularly on the 27th in Scotland and on the 29th to the 30th in southern England and East Anglia.

'December 1st was wet in extreme southern and south-west England and Wales and the 5th to the 7th was wet over most parts of England and Wales, with thunderstorms in the south and snow in the north. The periods from the 8th to the 15th and the 22nd to the 29th were mainly dry over England and Wales, although the 27th was generally snowy. The 16th to the 21st was generally wet, with considerable snowfall in the north.

'Over Scotland precipitation was mainly snowy throughout the month and occurred rather more frequently than over much of England and Wales. The last day of the year was very wet over the southern half of Britain.'

J. GRINDLEY

1977 The early part of the year followed a familiar pattern. The winter was not particularly cold but was prolonged well into spring. Owing to waterlogged fields, much spring sowing was dangerously late. An exceptionally hard frost in the second week of April did much damage to fruit blossom. Suddenly, as in the previous years, everything seemed to come right in May, which produced several weeks of brilliant sunshine, ending, unfortunately, just too early for the Jubilee celebrations.

Thereafter the weather veered away from the 1975/76 pattern. No prolonged summer drought developed, though there were sufficient dry spells for haymaking. Harvest prospects had never looked rosier, until heavy rains in August and early September came to dampen them down. It was a late harvest and, in the end, a costly one, with grain coming off the fields at high moisture contents and needing a lot of drying. Losses through shed and sprouted grain were fairly heavy, but little or no corn was abandoned.

With the late summer rains grass and fodder crops flourished amazingly. Never was there a greater abundance of autumn keep. And this exceptional growing season was marked by an unusual absence of frost. In many south-western districts no frost occurred until November 16th, a remarkably late date. This state of affairs greatly eased demands on hay and silage, enabling farmers to go into winter with their stocks virtually untouched. At the same time, there were sufficient dry spells in autumn to allow most of the planned sowing to be completed without difficulty.

Farm pests were less troublesome than usual. Early in summer fears of an aphid attack on wheat caused some crops to be sprayed aerially, but the menace did not develop. The same was not true of market garden crops, which were subjected to one of the worst onslaughts by black aphids ever known. Even crops normally immune, such as runner beans, marrows and pumpkins, were severely affected, though by autumn the infestation had passed and good late crops developed.

Both the extent (3,713,000 hectares) and the yields (nearly 17,000,000 tonnes of grain) showed an improvement on the harvest of the two previous years. Much of the grain, had, however, been badly damaged by rain, so while the price of cereals for livestock feed fell considerably that for grain of better quality (e.g. milling wheat and malting barley) remained fairly high.

Potatoes gave a much better crop than for several years. Sugar beet showed an increase over 1976 in both yields and sugar content. A feature of the previous few years, enhanced in 1977, was interest in oilseed rape. Between 1973 and 1977 the acreage grown increased from 14,000 hectares to 55,000 hectares.

Cattle numbers in the national herd remained fairly constant, a fall from 1,764,000 to 1,694,000 beef cows being roughly compensated for by an increase from 3,228,000 to 3,264,000 dairy cows. Average milk yields rose by 3·5%.

Sheep and poultry numbers and production remained

steady, but pigs, still harassed by low profitability, took a further knock. The total number of pig breeding stock fell by 7%. Towards the end of the year, however, the decline in the prices of coarse grain gave pig producers cause for a little more optimism.

Ministry of Agriculture statistics showed that the average earnings of men employed wholetime in agriculture were £50.27 in 1976, an increase of 15.9% over 1975 in money terms (though only 0.5% in real terms). A far cry from the 30 shillings a week wage in the 1930s.

RAINFALL SHOWS THAT ENGLAND AND WALES HAD A TYPICAL YEAR FOR WEATHER

Reproduced by kind permission of The Times

'After the extremes of the two preceding years, the prolonged drought of 1975–76 and the extremely wet September and October of 1976 over England and Wales, rainfall of 1977 was much more that of a 'typical' year. Totals for the year were close to the long-period average over Britain although, as usual, there were marked departures from average in individual months and groups of months. The annual total for Northern Ireland was 6 per cent less than average.

'The first four months of the year were generally wet over the country as a whole. February and March particularly so and that, after the previous wet autumn, meant that water supply authorities faced the coming summer with far happier prospects than in 1976. It was the wettest February since 1950 over England and Wales and since 1923 over Northern Ireland. More remarkably, the seven months from September 1976 to March 1977 was the second wettest period in the 250 year series beginning in 1727 for England and Wales as a whole (September, 1876 to March, 1877, was wetter).

'The spring months, April to May, were rather dry over England but the summer was marred by a wet June and rather wet August over southern England. Late winter and spring (February to April) were wet in the north; over Scotland it was the third wettest such period in a 110-year series (February to April in 1876 and 1903 were wetter) and it was the third wettest such period in a 79-year series over Northern Ireland (1940 and 1966 were wetter).

'The summer four months (May to August) were drier in

northern Britain than in the south, particularly so in Northern Ireland, where the dry series of months extended into October; it was the fourth driest May to September over Northern Ireland in the long series (May to September, 1911, 1933 and 1959 were drier). July was a dry month over practically the whole of the United Kingdom, especially England and Wales, where less than a tenth of the monthly average was recorded in the south Midlands and north-east England. Autumn (September to November) was exceptionally wet over Scotland, where it was the fifth wettest such period in the series (the most recent wetter autumn occurred in 1954).

'Among the largest and smallest rainfall totals recorded in 1977 were 4,586 mm at Llydaw, in Snowdonia, and 429 mm at Southend.

'January was unsettled and often snowy until its last few days. A cold spell occurred between the 9th and 19th and possibly the heaviest and most widespread snowfall of the year, at least over England, was experienced on the 12th to 13th. Flooding also followed heavy rain on the 25th.

'February, too, was thoroughly unsettled throughout, with prolonged and heavy rainfall at times. Flooding sometimes followed the rainfall and was particularly evident on the 9th, one of the wettest days of the year over the United Kingdom, and again after heavy precipitation, including snow, on the 22nd to 24th, when a front became almost stationary over the English Midlands. Generally mild weather was interrupted by a much colder spell on the 11th to 13th, when widespread snow fell on higher ground.

'The first general dry spell of the year was experienced from March 2nd to 9th over much of England and for a few days longer in eastern England; weather in Scotland was more unsettled. The period from the 10th to 23rd was generally wet, with a sharp transition from mild to colder weather on the 19th. Thunderstorms occurred in southern Britain on the 20th. Some snow, mainly light, was reported widely from the 26th to 29th, but the 30th was one of the wettest days of the year over Scotland.

'The first two days of April were also rather wet over Scotland and the period from the 3rd to the 10th was cold and snowy generally over Great Britain, although without any very heavy precipitation. The 13th to 18th was mainly dry as far north as the Central Lowlands but the last fortnight was more unsettled, with widespread thunderstorms on the 29th and 30th.

'The mainly unsettled weather that had persisted for so many months continued in the first half of May. Rainfall was often heavy. An extended dry spell began on the 14th and little or no rain fell over an area extending from mid-England and Wales to the extreme north of Scotland for almost three weeks. Heavy rain was experienced in south-west England on the 14th and 15th. There were widespread thunderstorms in southern England on the 25th and further isolated but heavy thunderstorms in western England and Wales on the 29th. The mainly dry weather of the latter half of May persisted everywhere in the United Kingdom, except Shetland, for the first three days of June.

'June 4th to 8th was rainy but not excessively so over most of the country (although the 6th was a very wet day in southern Scotland and northern England). Weather from the 9th was dominated by depressions over southern England or the Channel and the period up to the 13th (14th in the Midlands) was marked by extraordinarily wet weather with heavy thunderstorms over southern Britain. The wettest days were the 10th and 13th. The period from the 14th to 23rd was mainly dry everywhere and most of Scotland was practically rainless for 10 to 12 days. Some rain fell in the last week but amounts were not generally great except on the 28th, when heavy, short-period thunderstorms with large hail extended in a band from east Devon through the Midlands to the Wash, and again on the 30th.

'The period from mid-June to mid-August, although not entirely rainless, was the driest and most settled period of the year over England and Wales; in all, about 50 mm fell over the two countries, about a third of the average for the period. Scotland was more generally wet from July 15th to 24th and again from August 2nd to 4th, but the subsequent dry period extended up to August 23rd over that country.

'Isolated thunderstorms occurred in the south on July 8th, 11th and 12th. On the 8th, 20 mm was recorded in 16 minutes at Penmaen, where tornadoes were noted, and on the 12th, 80 mm fell in a short period at Weymouth, but the first fortnight was rainless at many places in Britain. The 17th was a wet day generally, and the next fortnight was unsettled, with heavy rain in northern Scotland. Even so, there were many rainless days and rainfall amounts in general were not large.

'The first few days of August were dry over most of England and Wales but weather over Scotland and Northern Ireland was more unsettled and heavy rain fell in those countries and in Cumbria and North Wales on August 4th; at Honister Pass in the Lake District, 140 mm was recorded on the 4th. Rainfall spread to southern England on the 5th and the next few days were unsettled there; more than 50 mm was recorded in thunderstorms in East Anglia on the 8th. Northern England, Scotland and North Wales remained mainly dry from the 5th to the 23rd and the 7th to 15th was generally dry over southern England. A period of remarkably wet weather began in southern Britain on the 16th with heavy thunderstorms occurring from place to place on many days. On the 16th more than 110 mm was recorded in the western suburbs of London (Ruislip–Hillingdon) and in neighbouring Buckinghamshire and Hertfordshire.

'Further heavy thunderstorms followed on the 17th and again 50 mm was exceeded over a wide area extending from Somerset to east Wales and the Midlands. Thunderstorms were widespread and heavy again on the 21st and more than 70 mm was recorded in east Devon on the 23rd. Rain spread to the whole of the United Kingdom on the 24th, bringing the dry spell in northern Britain to an end; it was probably the wettest day of the year over the United Kingdom as a whole. Further heavy storms occurred in North Wales and Lancashire on the 25th and north-west England, the Borders and Lothian on the 26th. It was the turn of some south-east London suburbs to experience heavy thunderstorms on the 27th. The Bank Holiday, August 28th, was dry, warm and sunny but the last few days were again unsettled, with heavy rain in southern Scotland on the 30th.

'The first nine days of September and the last week were unsettled and wet, particularly so over Scotland, where generally heavy rain fell on the 5th, 9th and the last few days. Mid-month, September 9th to 23rd, over southern Britain and 12th to 24th over northern Britain and Northern Ireland, was generally settled and dry. Although heavy rainfall occurred on many days over Scotland, amounts over England and Wales, which were more directly affected by high pressure over and to the south of the country, were generally small.

'The first 11 days of October were mainly unsettled and wet. Rainfall on the 3rd approached 150 mm in the hills of Gwynedd and more than 50 mm was recorded in north-west England, the Isle of Man and southern Scotland. Further heavy rain fell on the 5th in hilly areas extending from Cornwall (where 99 mm was recorded at Goonhilly) to Strathclyde. Extensive flooding occurred on the 7th in Glasgow, when heavy rain fell in the Central Lowlands. Much of England and Wales experienced a dry, settled period from the 10th to 19th, and Scotland was mainly dry in the week beginning on the 12th. Heavy thunderstorms occurred in south-east England and East Anglia in an unstable southerly airstream on the 20th; much flooding was reported, especially in London. On the 23rd amounts exceeded 125 mm in Cumbria and Gwynedd and more than 50 mm was recorded in parts of Lancashire and south-west Scotland.

'October 25th to 28th was mainly dry over England, Wales and southern Scotland, but perhaps the most notable rainfall event of the year occurred on the night of October 30th–31st when extremely heavy rainfall, accompanied by gales, was recorded over an area extending from South Wales to Scotland, including much of the Pennines. More than 100 mm was measured for the rainfall day, the 30th, in counties as far apart as Dyfed and Lothian and the heaviest daily point rainfall of the year, 182 mm, was reported at Thirlmere, in the Lake District. Extensive flooding followed the rainfall, particularly in south-west Scotland; on the 31st the only road and rail links between England and Scotland were in the extreme east.

'The wet weather of late October continued for the first two days of November; November 1st was particularly wet in southern Britain and many amounts exceeding 50 mm were reported in Wales. The first three weeks of November remained generally unsettled. Thunderstorms occurred from the 5th to 8th and the period from the 14th to the 24th was snowy, particularly in Scotland, after a change from very mild to cold weather on the 13th. The last week was mainly dry.

'The dry weather of late November continued for the first five days of December, except in extreme south-west

England. The dry start was followed by a week of wet, unsettled weather, particularly in southern Britain on the 7th and 9th; the 7th was snowy in the Midlands and northern England and southern Scotland. The period 13th to 20th was mainly dry (apart from general rainfall on the 17th in the south), but the next week was much more unsettled with very heavy rainfall from the 22nd to 24th in hilly districts from North Wales to western Scotland. The last few days of the year were mainly dry but there was some snowfall over northern Scotland.'

J. GRINDLEY

Appendix A

ANNUAL RAINFALL TABLES

The tables which follow are reproduced by kind permission of The Times.

1947 RAINFALL

General values of rainfall are given in the following table:

	1947	*Average,* 1881–1915	*Difference from average*	*Per cent. of average*
	Inch	Inch	Inch	
England and Wales	32·7	35·2	—2·5	93
Scotland	48·7	50·3	—1·6	97
Northern Ireland	41·6	39·7	+1·9	105

1947 Monthly rainfall is shown in the following table in inches and as a percentage of average for the month:

	England and Wales		Scotland		Northern Ireland	
	In.	%	In.	%	In.	%
January	3·4	113	5·7	115	3·5	105
February	2·1	83	1·8	43	1·2	39
March	6·8	255	4·0	106	4·8	156
April	3·0	143	6·4	215	4·0	156
May	2·4	103	3·7	124	4·2	156
June	2·7	112	4·2	147	4·5	159
July	2·7	95	3·8	99	4·0	118
August	0·5	16	0·2	4	0·6	14
September	2·0	78	5·4	135	3·3	110
October	0·8	20	2·7	56	2·2	60
November	3·0	86	7·3	138	5·8	153
December	3·3	83	3·5	59	3·5	83

1947 Annual totals for some representative stations are as follows. The first figure indicates the rainfall in inches and the second the percentage of average:

London	20·97	86
Margate	18·99	83
Ventnor	28·49	99
Lowestoft	17·16	73
Sandringham	19·37	73
Princetown	79·48	97
Bath	26·98	89
Malvern	24·77	90
Manchester	26·49	84
York	24·25	100
Middlesborough	21·08	89
Seathwaite	106·50	87
Abergavenny	37·09	99
Rhayader	38·78	83
Holyhead	33·67	96
Douglas	43·94	107
Dumfries	43·30	111
Edinburgh	25·92	99
Glasgow	38·63	107
Tiree	39·05	87
Balmoral	31·32	95
Aberdeen	26·26	89
Fort William	69·72	88
Glen Quoich	97·79	88
Ullapool	34·61	74
Wick	26·66	89
Belfast	39·09	113
Londonderry	43·68	105

1957 RAINFALL

General values of rainfall are given in the following table:

	1957	*Average* 1881–1915	*Difference from average*	*Per cent of average*
	In.	In.	In.	
ENGLAND	32·5	32·7	−0·2	99
WALES	55·4	50·1	+5·3	111
ENGLAND AND WALES	35·7	35·2	+0·5	101
SCOTLAND	53·6	50·3	+3·3	107
NORTHERN IRELAND	43·0	39·7	+3·3	108

1957 Monthly rainfall is shown in the following table in inches and as a percentage of average for the month:

	ENGLAND AND WALES		SCOTLAND		NORTHERN IRELAND	
	In.	%	In.	%	In.	%
January	2·9	97	7·2	147	5·2	156
February	3·9	153	4·3	103	2·7	92
March	2·7	102	4·7	116	3·3	108
April	0·4	19	2·1	68	1·5	57
May	1·9	83	2·2	74	2·1	80
June	1·9	79	2·8	99	2·1	73
July	4·1	144	5·7	152	4·9	144
August	4·1	121	5·7	127	4·7	114
September	5·0	195	4·0	99	5·1	170
October	3·0	74	5·3	109	4·3	118
November	2·5	72	2·9	56	1·6	41
December	3·3	83	6·7	114	5·5	132

1957 Annual totals in inches and as a percentage of average are given for representative stations in the following table:

	In.	*Per cent of average* 1916–50
London	22·35	91
Dover	29·41	104
Worthing	25·30	92
Sandringham	26·00	98
Marlborough	31·69	102
Portland Bill	27·25	108
Princetown	87·89	107
Falmouth	41·99	96
Birmingham	29·09	100
Bolton	48·09	112
Bradford	32·26	95
Barnard Castle	30·67	101
Durham	21·64	88
Keswick	56·36	103
Ystalyfera	72·90	112
Aberystwyth	39·18	107
Rhyl	26·60	103
Douglas	43·92	107
Eskdalemuir	59·74	106
Edinburgh	25·33	95
Kilmarnock	39·92	107
Paisley	41·84	103
Inverary	84·44	110
Perth	32·16	104
Balmoral	31·87	96
Nairn	27·20	114
Fort William	82·94	105
Skye	59·08	110
Lochinver	57·03	128
Orkney	41·80	131
Armagh	35·92	113
Londonderry	42·29	101

1967 RAINFALL

General values of rainfall are given in the following table:

	1967	*Average 1916–50*	*Difference from average*	*Per cent. of average*
	In.	In.	In.	
ENGLAND	35·2	32·7	+2·5	108
WALES	61·1	53·4	+7·7	114
ENGLAND AND WALES	38·7	35·6	+3·1	109
SCOTLAND	59·3	55·9	+3·4	106
NORTHERN IRELAND	45·2	42·6	+2·6	106

1967 Monthly rainfall is shown in the following table in inches and as a percentage of average for the month:

	ENGLAND AND WALES		SCOTLAND		NORTHERN IRELAND	
	In.	%	In.	%	In.	%
January	2·3	64	4·9	81	3·2	74
February	3·5	133	5·5	131	3·4	113
March	2·1	95	6·9	198	3·4	131
April	1·9	79	2·6	75	2·1	79
May	5·4	219	5·1	149	4·7	166
June	1·6	75	2·6	77	1·6	56
July	2·5	79	3·8	86	3·4	90
August	3·0	94	3·8	79	3·8	96
September	3·8	128	5·5	109	5·5	145
October	6·5	179	9·2	149	6·3	143
November	3·0	79	4·8	85	3·7	90
December	3·1	88	4·6	81	4·1	94

1967 Annual totals in inches and as a percentage of average are given for representative stations in the following table:

	In.	*Per cent of average* 1916–50
London (St. James's)	26·38	113
Herne Bay	21·84	96
Worthing	29·16	106
Bristol	34·38	111
Weymouth	34·92	116
Plymouth	36·76	97
Newquay	35·67	104
Birmingham	30·17	98
Bradford	38·34	114
Hull	24·43	96
Durham	28·26	107
Norwich	22·30	88
Cardiff	49·66	118
Rhyl	29·99	113
Aberporth	37·96	93
Llandrindod Wells	46·14	114
Douglas	50·55	112
Eskdalemuir	66·04	106
Edinburgh	25·08	91
Prestwick	36·76	100
Glasgow	42·69	107
Islay (Eallabus)	57·06	108
Crieff	35·61	86
Braemar	37·53	103
Inverness	23·76	91
Glenleven (Blackwater Dam)	130·03	136
Skye (Duntulm)	62·11	117
Stornoway	51·86	131
Wick	34·26	114
Lerwick	58·51	131
Armagh	35·55	107
Londonderry	50·36	109
Belfast	35·84	95

1977 RAINFALL

General values of rainfall are given in the following table:

	1977	*Average 1941–70*	*Difference from average*	*Per cent of average*
	mm.	mm.	mm.	
ENGLAND	848	837	+11	101
WALES	1413	1385	+28	102
ENGLAND AND WALES	925	912	+13	101
SCOTLAND	1457	1431	+26	102
NORTHERN IRELAND	1026	1095	−67	94

1977 Monthly rainfall is shown in the following table in mm. and as a percentage of average for the month:

	ENGLAND AND WALES		SCOTLAND		NORTHERN IRELAND	
	mm.	%	mm.	%	mm.	%
January	101	117	127	93	104	100
February	138	212	137	132	152	203
March	73	124	129	140	82	117
April	51	88	116	129	74	109
May	52	78	74	81	26	36
June	85	139	78	85	53	67
July	24	33	63	56	49	53
August	102	113	110	85	98	95
September	36	43	146	107	69	64
October	64	77	178	119	105	98
November	104	107	207	146	117	115
December	95	106	92	60	97	85

1977 Annual totals in mm. and as a percentage of average are given for representative stations in the following table:

	mm.	*Per cent. of average* 1941–70
London (St James's)	636	104
Margate	524	91
Worthing	787	110
Bristol	828	106
Cambridge	540	97
Plymouth	1034	104
Penzance	1305	119
Birmingham	784	101
Carlisle	782	90
Bradford	887	102
Hull	634	98
Durham	594	91
Lowestoft	474	79
Cardiff	984	89
Colwyn Bay	923	123
Aberporth	893	94
Builth Wells	1010	99
Douglas	1188	104
Eskdalemuir	1693	112
Edinburgh	777	115
Prestwick	959	105
Glasgow	1141	109
Islay (Eallabus)	1295	101
Pitlochry	784	89
Braemar	1019	116
Inverness	572	93
Montrose	743	102
Skye (Prabost)	1410	78
Stornoway	1103	101
Wick	753	96
Lerwick	1153	98
Armagh	789	91
Londonderry	912	96
Belfast	1063	96

Appendix B

TABLE OF COMPARATIVE PRICES 1257–1977

	Wheat (per qtr)		Oats (per qtr)		Barley (per qtr)		Rye (per qtr)		Malt (per qtr)		Peas (per qtr)	
	s	d	s	d	s	d	s	d	s	d	s	d
1257	24	0										
1258	17	0										
1259	5	9¾	1	8	3	5½	2	0				
1260	4	9	1	6½	3	5			3	5		
1261	4	3	2	0¼								
1262	6	1	1	8¼	3	2	3	8⅞				
1263	3	11⅜	1	8	3	6¾					3	0
1264	4	4	2	2	4	0					3	4½
1265	3	3	1	4⅞							2	0
1266	4	5½			4	0						
1267	4	5¼	1	5¼	2	5¾						
1268	5	3⅜	2	7½	3	6¾						
1269	5	0	1	7¼	3	1						
1270	6	4⅛			4	0½						
1271	6	10⅞			5	1½						
1272	6	4⅝	2	4⅝	4	7					4	3¼
1273	5	5¾	2	4⅛	4	0½						
1274	6	9⅛	2	7¾	4	8⅞						
1275	5	0⅞	2	2½	3	10⅞						
1276	6	2⅝	2	8⅝	4	10						
1277	5	1¾	2	3⅝	4	1					4	0¾
1278	4	4⅝	2	4⅜	3	8⅞					2	9⅜
1279	5	1¼	2	0⅝	3	11⅜						
1280	4	11⅞	2	4⅜	3	6¾						
1281	6	0¾	2	5	3	5⅜					4	4¼
1282	5	11½	2	1½	4	1¼					3	2½
1283	6	11¼	2	4¼	4	5⅛					4	9⅝
1284	4	11¾	1	9⅞	3	1⅝					2	10¾
1285	5	4¼	2	1¾	3	6⅛	4	1⅝	4	0	3	3¾
1286	4	9	2	0⅞	3	3⅛					3	1¾
1287	2	10¼	1	5¾	2	6½					1	11¾
1288	3	0⅞	1	6½	2	3⅝					2	0¾
1289	4	3⅝	1	11¾	3	2½					2	6¼
1290	6	5½	2	6¾	4	5⅝					4	0
1291	5	7¼	2	2⅞	4	4½	4	3¼	5	1	4	6
1292	5	4⅝	2	4⅝	3	11⅝					3	6
1293	8	3⅛	2	9½	5	1					5	2¾
1294	9	1⅛	2	10¼	6	1⅜					7	8¾
1295	6	9	2	4¾	4	4⅜					4	7⅞
1296	4	9¼	2	3¼	3	9⅛	3	9⅜	4	7¾	3	7⅜
1297	5	2½	2	4¾	4	2⅞						

Vetches (per qtr)		Beans (per qtr)		Wool (per lb)		Horses		Cattle		Sheep		Pigs		Hens	
s	d	s	d	s	d	s	d	s	d	s	d	s	d	s	d
					$4\frac{3}{4}$	5	6	6	11		9	2	0		1
3	0														
		2	6												
		2	0		8										
					$6\frac{1}{2}$										
		2	3												
		4	$6\frac{3}{4}$		$2\frac{3}{4}$										
					3	10s to 14s		6s to 12s							
5	10				$3\frac{3}{8}$										
3	4	4	$7\frac{1}{2}$		3										
		3	6												
		4	$9\frac{3}{8}$												
		3	$3\frac{7}{8}$												
		4	$6\frac{3}{4}$	10s 11d per tod											
3	$6\frac{1}{8}$	5	6		9										
3	$9\frac{1}{8}$	3	$2\frac{3}{8}$												
					3										
2	7														
		5	$0\frac{5}{8}$		4										
		4	$1\frac{1}{8}$												
		4	$7\frac{1}{2}$		$3\frac{1}{2}$										
		2	$9\frac{3}{8}$												
2	$9\frac{1}{8}$	3	$6\frac{7}{8}$												
		3	$0\frac{3}{4}$												
		1	$9\frac{3}{4}$												
		2	$2\frac{1}{4}$												
		2	$6\frac{3}{4}$												
		4	4												
4	7	4	$8\frac{3}{4}$												
		5	0												
		4	$11\frac{1}{8}$												
		6	$3\frac{1}{2}$			15s/17s		5s/10s			6	3d/6d			
		5	$1\frac{7}{8}$												
3	$3\frac{1}{2}$	3	$7\frac{3}{4}$												
		4	4												

	Wheat (per qtr)		Oats (per qtr)		Barley (per qtr)		Rye (per qtr)		Malt (per qtr)		Peas (per qtr)	
	s	d	s	d	s	d	s	d	s	d	s	d
1298	5	2⅛	2	5½	4	3⅝					3	5¼
1299	5	9¾	2	9⅛	4	4½					2	7¾
1300	4	9	1	11⅜	3	8½					2	4¼
1301	5	0⅛	2	0⅝	3	7⅝						
1302	4	11⅞	2	1⅜	3	4⅞					3	0¾
1303	4	1¼	2	1⅞	2	10⅜					3	1⅛
1304	5	9⅞	2	4¾	4	1¼					3	10⅞
1305	4	10⅞	2	9¾	3	10⅜					4	6
1306	3	11⅜	2	1⅜	3	5½					4	5⅜
1307	5	6½	2	4⅛	3	8¾					3	0¼
1308	6	11¼	2	8¾	4	4⅞					4	4¼
1309	7	9⅜	3	3	5	2					5	2⅝
1310	7	0½	2	7⅞	5	1					3	7½
1311	4	5¼	2	5⅝	3	9					2	10⅝
1312	4	11⅜	2	4	3	11	3	6	3	5	2	9
1313	5	6⅜	2	8¼	4	0⅜					3	3⅛
1314	8	4¾	2	8¾	5	4					4	1¼
1315	14	10⅞	4	10½	13	1	11	11⅝	13	0	11	2½
1316	15	11⅞	5	4⅞	8	9⅝					11	7¼
1317	8	3½	3	9½	5	7					5	8
1318	4	6½	2	1	3	5½					3	6⅜
1319	5	9⅝	2	2¾	3	5⅜					3	4¼
1320	6	5	2	5	4	1¾					3	9¼
1321	11	7¾	4	0¾	8	5					9	3⅝
1322	8	11⅞	3	2½	6	6					6	4¾
1323	7	5⅜	2	7¼	4	4⅛					5	1¼
1324	7	4⅝	2	6½	5	4¾					5	1⅝
1325	5	8⅜	2	1	3	8½					3	7
1326	3	7⅞	1	11⅜	3	0½					2	9½
1327	3	11	2	0⅛	2	10⅝					2	6¾
1328	6	5½	3	1	4	5¾					4	11⅝
1329	6	6⅝	2	5¾	4	6⅛					3	6⅜
1330	7	2¼	2	11⅛	5	2¾					3	11¼
1331	7	11¼	3	3½	6	3⅛	5	9½	6	2⅝	5	5¾
1332	4	8⅝	2	2	3	6					3	5½
1333	4	2⅜	2	2⅛	3	3⅝					3	0⅝
1334	4	0⅛	1	10¾	2	10¾					3	0⅞
1335	5	3½	2	2⅜	3	9⅞					2	10¼
1336	4	11	2	1⅛	3	8½					3	1
1337	3	7	1	7½	2	7⅜					2	0⅝
1338	3	4	1	6	2	1					1	8½
1339	5	11¾	1	8⅞	3	1⅛					2	9¾
1340	3	6½	1	11	2	9½					2	5¼

Vetches (per qtr)		Beans (per qtr)		Wool (per lb)		Cattle	Sheep
s	d	s	d	s	d	s d	
		4	3¼				
		2	3⅞				
		2	8				
		2	8		4		
		3	11⅝				
		6	3				
		3					
		3	8½				
		4	11				
		7	4				
		4	3½				
		3	1⅝				
2	5⅜	2	11				
		3	4⅝			£2.12.0	5 0
		5	4¼				
13	0	12	8				
		13	0⅞				
		5	10⅞				
		3	10⅞				
		3	4				
		4	1⅝		9½		
		9	11⅞				
		7	8				
		5	3				
		5	4½				
		3	11				
		4	0				
		3	9				
		5	7				
		4	0				
		6	0				
		6	4⅜				
		3	11				
3	4	3	4½				
		3	6				
2	11⅞	3	4⅞				
2	10⅝	2	10				
		3	4				
		1	11⅛				
		3	0½				
		2	11¼				

	Wheat (per qtr)		Oats (per qtr)		Barley (per qtr)		Peas (per qtr)		Vetches (per qtr)		Beans (per qtr)	
	s	d	s	d	s	d	s	d	s	d	s	d
1341	3	$9\frac{5}{8}$	1	$10\frac{3}{8}$	3	$0\frac{5}{8}$	2	$1\frac{1}{2}$			2	$6\frac{1}{2}$
1342	4	$1\frac{5}{8}$	2	$0\frac{3}{4}$	3	$2\frac{3}{8}$	2	10			2	$9\frac{3}{8}$
1343	5	$7\frac{3}{4}$	2	$1\frac{3}{8}$	3	$8\frac{5}{8}$	3	$1\frac{1}{2}$			3	$10\frac{3}{8}$
1344	3	6	1	$9\frac{1}{2}$	2	$9\frac{3}{4}$	2	5			2	$11\frac{1}{2}$
1345	3	$9\frac{7}{8}$	2	$0\frac{1}{4}$	2	$9\frac{3}{4}$	2	$3\frac{3}{4}$			3	0
1346	6	$10\frac{1}{2}$	2	$7\frac{3}{4}$	3	$11\frac{7}{8}$	3	$7\frac{7}{8}$			5	5
1347	6	$7\frac{3}{8}$	2	$4\frac{7}{8}$	4	$9\frac{1}{4}$	3	$2\frac{3}{8}$			3	$6\frac{3}{8}$
1348	4	2	1	$5\frac{3}{8}$	2	$6\frac{1}{2}$	2	3			3	$3\frac{1}{2}$
1349	5	$5\frac{7}{8}$	2	$6\frac{1}{4}$	3	$10\frac{1}{2}$	3	$4\frac{1}{4}$			4	5
1350	8	$3\frac{1}{8}$	3	8	6	4	4	5			4	$11\frac{1}{8}$
1351	10	$2\frac{1}{2}$	3	$7\frac{3}{4}$	6	$10\frac{1}{4}$	6	$0\frac{1}{4}$			6	1
1352	7	$2\frac{1}{8}$	4	$0\frac{3}{8}$	5	$10\frac{1}{4}$	6	$4\frac{3}{4}$			6	$9\frac{3}{8}$
1353	4	$2\frac{1}{2}$	2	$3\frac{7}{8}$	3	$0\frac{5}{8}$	2	$3\frac{1}{8}$			2	$7\frac{1}{8}$
1354	5	$3\frac{3}{4}$	2	$0\frac{3}{4}$	3	5	2	$6\frac{3}{4}$			3	$0\frac{1}{2}$
1355	5	$11\frac{3}{8}$	2	$9\frac{1}{4}$	3	10	3	0			3	4
1356	6	0	2	$10\frac{5}{8}$	4	5	5	$3\frac{3}{8}$	5	$9\frac{3}{4}$	5	$6\frac{3}{8}$
1357	6	$10\frac{1}{4}$	2	10	4	$5\frac{1}{8}$	4	$0\frac{3}{8}$	4	$2\frac{3}{8}$	4	$2\frac{3}{4}$
1358	5	$6\frac{1}{2}$	2	$7\frac{1}{8}$	5	$1\frac{1}{2}$	3	$2\frac{1}{8}$	3	$8\frac{3}{8}$	2	$9\frac{5}{8}$
1359	5	11	2	$4\frac{3}{8}$	4	$4\frac{1}{4}$	2	$6\frac{3}{8}$	2	$6\frac{5}{8}$	3	10
1360	6	$3\frac{1}{2}$	2	$9\frac{1}{4}$	4	$6\frac{1}{2}$	4	$1\frac{5}{8}$			5	$2\frac{3}{8}$
1361	5	$4\frac{3}{4}$	3	$2\frac{1}{2}$	4	$7\frac{1}{4}$	4	$9\frac{3}{8}$	4	$8\frac{1}{2}$	5	6
1362	7	6	3	6	5	$5\frac{1}{4}$	5	$7\frac{1}{2}$	6	8	6	8
1363	8	6	2	$10\frac{5}{8}$	5	$2\frac{5}{8}$	3	$8\frac{3}{8}$			4	$11\frac{3}{4}$
1364	7	$5\frac{3}{8}$	2	$8\frac{5}{8}$	4	$2\frac{1}{4}$	3	$5\frac{1}{2}$			5	0
1365	6	$0\frac{3}{8}$	2	$5\frac{3}{4}$	4	$2\frac{5}{8}$	3	$3\frac{1}{2}$	3	$4\frac{3}{4}$	3	8
1366	6	$8\frac{1}{2}$	2	$11\frac{3}{4}$	4	$9\frac{1}{8}$	3	$0\frac{1}{2}$			4	0
1367	8	$7\frac{1}{2}$	3	$1\frac{1}{2}$	4	$1\frac{5}{8}$	3	7	5	0	3	4
1368	3	$7\frac{5}{8}$	3	$0\frac{5}{8}$	4	$7\frac{1}{2}$	4	$5\frac{3}{8}$	3	$9\frac{3}{8}$	5	$6\frac{3}{8}$
1369	11	$10\frac{1}{4}$	4	$2\frac{5}{8}$	8	$5\frac{1}{4}$	6	8	5	9	7	$10\frac{3}{8}$
1370	9	$4\frac{5}{8}$	3	$6\frac{7}{8}$	4	6	5	1	3	8	6	$3\frac{1}{4}$
1371	6	$11\frac{3}{8}$	2	$3\frac{1}{4}$	4	$1\frac{7}{8}$	3	$5\frac{3}{4}$	3	0	4	$6\frac{3}{8}$
1372	7	$10\frac{1}{4}$	2	$9\frac{3}{8}$	4	$9\frac{7}{8}$	3	$4\frac{1}{2}$	3	$2\frac{1}{4}$	4	$8\frac{1}{4}$
1373	6	$2\frac{1}{4}$	2	$6\frac{1}{4}$	3	$10\frac{5}{8}$	3	7			4	$3\frac{3}{8}$
1374	8	$2\frac{1}{4}$	2	$9\frac{3}{4}$	4	$10\frac{1}{4}$	4	$0\frac{3}{4}$			5	$3\frac{1}{4}$
1375	7	$9\frac{1}{8}$	2	$9\frac{3}{4}$	5	0	4	$0\frac{1}{8}$			4	$10\frac{3}{8}$
1376	4	9	2	$3\frac{3}{8}$	3	$9\frac{1}{4}$	3	$0\frac{7}{8}$			4	$1\frac{1}{2}$
1377	3	$8\frac{3}{8}$	2	4	3	$1\frac{7}{8}$	2	$4\frac{5}{8}$	2	0	3	6
1378	3	$6\frac{7}{8}$	1	$11\frac{3}{4}$	2	$6\frac{1}{2}$	2	$7\frac{3}{4}$			3	4
1379	5	$9\frac{3}{8}$	1	$11\frac{5}{8}$	2	9	2	$10\frac{1}{8}$			3	11
1380	6	$2\frac{7}{8}$	2	4	3	$7\frac{3}{8}$	3	$3\frac{1}{2}$			3	7
1381	5	$7\frac{1}{4}$	2	$2\frac{3}{8}$	3	4	2	$11\frac{3}{8}$			5	6
1382	5	$3\frac{1}{2}$	1	$11\frac{1}{2}$	3	$0\frac{5}{8}$	2	$8\frac{3}{8}$	3	4		
1383	4	10	2	$3\frac{5}{8}$	3	$11\frac{3}{8}$					3	$5\frac{3}{4}$

	Wheat (per qtr)		Oats (per qtr)		Barley (per qtr)		Rye (per qtr)		Malt (per qtr)		Peas (per qtr)		Beans (per qtr)	
	s	d	s	d	s	d	s	d	s	d	s	d	s	d
1384	5	7	2	2⅝	3	1					3	3⅛	4	8
1385	5	0⅝	2	5⅝	3	2½					3	4	5	0
1386	4	1	1	6	2	11½					2	2⅝	4	6
1387	3	4¾	1	4⅜	2	8⅛					1	9¾		
1388	3	8⅛	1	11	2	10⅜					2	0¼	3	0
1389	5	5⅜	2	2	3	0⅜					2	8⅜	4	0
1390	8	9	3	7	5	8¼					4	8½		
1391	5	5⅞	2	3¼	3	4⅞					3	3½	4	6⅛
1392	3	2⅝	1	10¼	2	4⅜	2	2			2	10		
1393	3	8¾	1	11⅞	2	8⅛					3	0⅞		
1394	3	10¾	2	1⅝	3	2⅜					2	11¼	4	10
1395	5	0	2	4⅜	3	2¼	2	9			3	4¾	3	8
1396	5	11½	2	7½	3	3⅝	4	2⅝			3	4⅜		
1397	5	9⅝	3	4	4	4⅝					5	5		
1398	5	2¾	2	2	3	4⅝					2	6½	4	0
1399	5	6⅜	2	2¾	3	6¾					2	11		
1400	7	11⅛	2	3¼	6	3⅝					4	6	3	10⅝
1401	7	5¾	2	7	4	6½	6	5¼	5	4			3	1⅜
1402	6	8¾	2	3	4	4½	5	8	4	9¼			4	6
1403	4	11¼	2	2½	4	2⅜	3	9¾	3	6			3	9¼
1404	4	0	2	2	3	4¾	2	8½			2	8¾	2	0
1405	3	9¾	1	11½	2	8½	2	5	3	3¼			3	4
1406	4	4	1	10¼	2	11	3	2	3	6¾			2	5
1407	4	6¾	2	1½	3	7	3	7	4	10			3	0
1408	7	3¼	2	10	4	4¾	4	1½	5	1¾			2	6
1409	8	11½	3	2½	5	5	6	4	4	8½	4	8	4	3
1410	4	10½	2	3	4	2	4	1	4	0½				
1411	4	10	2	1	3	3¼	2	9	3	11½			3	6
1412	4	10¾	2	0¾	2	11½	3	4¾	4	3¾			3	1¼
1413	4	3½	2	0¾	3	3¼	3	4¾	4	3¾				
1414	4	3¾	2	2¾	3	8¼	2	11	5	0¼			3	0½
1415	6	3½	2	10	4	6½	4	0½	5	1½	4	11¼		
1416	7	11¾	2	2¾	4	6	3	6	5	7¾	3	4		
1417	5	3½	2	2	3	8¼	4	8½	4	6				
1418	6	11½	2	0¼	3	7	3	0½	3	3			2	8
1419	4	9¼	2	3¼	4	2¾	2	9½	4	2½	3	6½	3	4
1420	6	3	2	1½	3	4	3	4	4	1½	3	3		
1421	5	2¾	2	5¼	3	4¼	3	11	3	7½	2	11		
1422	4	4	2	2¼	3	8¾	3	0½						
1423	4	5½	1	11¼	3	5	3	6	4	3¼	2	10	4	0
1424	4	11¼	1	11¼	3	3½	3	10¾	4	6¼				
1425	4	0¼	1	10¼	3	1½	3	0	4	5½	2	4	4	0
1426	3	11¼	1	11½	3	3¾	3	6	3	6¼			3	0

	Wheat (per qtr)		Oats (per qtr)		Barley (per qtr)		Rye (per qtr)		Malt (per qtr)		Peas (per qtr)		Beans (per qtr)	
	s	d	s	d	s	d	s	d	s	d	s	d	s	d
1427	4	4	2	2	3	1¾	2	11	3	5			4	0
1428	8	10¾	2	8½	4	8¾	5	5	5	5			4	0
1429	7	11	2	6½	4	4½	7	4	7	2			4	0
1430	5	11¼	2	0½	3	4¼	4	10	5	6			3	4
1431	4	8	1	9½	3	1	4	0	3	11½	3	0		
1432	6	11	2	5	3	8½	4	8	4	3½			6	8
1433	5	10¼	2	3	4	2¾	5	4	4	6½			4	0
1434	5	4½	1	11½	2	10	3	5	3	4½	3	5¾		
1435	5	6¼	1	8½	2	5	3	5¼	2	8			4	0
1436	5	5½	1	10¾	3	0¾	3	11¼	3	5¾			4	0
1437	9	3¾	2	11¼	4	0	6	8	5	5½			3	9
1438	14	7½	3	4¼	6	8¾	11	6	6	8¼			7	4½
1439	7	6¾	2	3	5	2¼	4	10½	7	2			4	0
1440	3	10½	1	7¼	3	0	3	0	2	11¾	2	6		
1441	4	0¼	1	10½	2	4½			2	6	2	3		
1442	3	11¼	2	1¼	3	2¾	2	8	3	7½	2	8		
1443	4	2	1	9	3	1½			3	3¼	3	5½		
1444	3	11¾	1	8½	2	5½	2	0	2	7¼			4	0
1445	6	3½	1	11	2	7½	2	10½	2	10¾				
1446	5	11¾	2	1¾	2	10½	4	0	3	5¾				
1447	5	2	1	11	3	2¾	3	9¼	3	7				
1448	5	7½	1	10	3	1½	4	0	3	9½			2	2
1449	5	3½	1	8½	2	10¼	4	0	3	6			2	1
1450	6	6¾	2	0	3	2½	5	4	4	3			3	11
1451	6	6	1	9¾	3	0	4	11	3	5½			3	4
1452	5	8¾	1	8	2	11½	4	2½	3	2			2	2
1453	5	1¼	2	2¾	3	6½	3	4½	4	5			4	0
1454	3	2¼	1	8½	2	9½			3	4¾	2	3¼		
1455	5	5½	1	10	3	1¼	3	0	4	0¾	2	7		
1456	4	11¾	2	0	2	10¾	3	4	3	1			3	0
1457	5	9½	1	10½	2	10½	3	9¾	3	3¾			2	11½
1458	5	9¼	1	9¼	3	6	4	0	2	8¾				
1459	5	1¾	1	10	3	1¼	3	4	3	1½			3	0
1460	7	0¼	2	0¾	4	0½	5	0	4	4				
1461	7	5¼	2	3¾	4	1			4	5¼	4	4		
1462	4	4½	1	5½	3	3½	4	8	2	4			3	3
1463	3	10½	1	8½	2	4½	2	10	3	0			3	6
1464	4	1½	3	4	3	6½	3	4	4	10¾			3	8
1465	4	7	1	9	3	0½	3	0	4	4¼			4	0
1466	5	4	2	1	3	4¾	3	3¼	3	8			3	4
1467	5	4	2	1½	3	2	4	0	3	9¾			3	4¼
1468	5	7¾	1	8½	3	2½	3	8	3	0			2	7½
1469	6	5½	2	1¼	3	6¼	4	2¾	3	0	2	8		

	Wheat (per qtr)		Oats (per qtr)		Barley (per qtr)		Rye (per qtr)		Malt (per qtr)		Peas (per qtr)		Beans (per qtr)	
	s	d	s	d	s	d	s	d	s	d	s	d	s	d
1470	5	9¾	1	9	3	4	5	4	3	8½				
1471	5	7½	1	10¼	3	11¼			4	9¼			4	0
1472	4	0¾	1	10¼	3	5			3	3¼			3	4
1473	3	10	2	5	3	6¾	3	4	3	1¼			3	4½
1474	4	6	1	5½	3	2¾			3	0½			2	8¾
1475	5	5¼	1	11¼	3	2	4	0	2	4			3	4
1476	5	1½	1	11½	3	2	3	4	2	4¼			3	5
1477	6	8	1	8	3	2¾			2	6¼			2	8
1478	6	7¼	1	10	2	9	4	0	4	3½			4	0
1479	5	10¼	1	7	3	4½	3	4	3	6¼			4	0
1480	5	10	1	11¼	3	0	3	4	4	3½			3	10¾
1481	8	6¾	3	3¼	5	8	5	0½	5	10¼			5	0
1482	10	4	2	4¾	6	2¾			7	0	8	4	5	9¼
1483	7	3¼	2	4	5	11			4	9	4	1	4	1¾
1484	5	3¾	2	2½	4	1¼	4	8	3	10¼			3	8
1485	4	6¼	1	8	3	5			2	4	3	4	3	4
1486	5	3½	1	9	4	11	6	8	3	4			3	6
1487	5	5¼	1	9¾	3	0½	5	4	3	10¼	2	0		
1488	5	6	2	9¾	4	0¼	4	0	2	10¾			2	0
1489	5	10¾	1	9½	3	3	6	8	3	6			3	5
1490	4	10½	1	8¾	4	2			4	0				
1491	6	7¼	2	0	3	7	4	4	3	3			3	0
1492	4	3	1	4	4	0	2	0	3	11				
1493	4	1	3	4	3	3			3	4				
1494	4	9¾	1	8¾	3	0			3	3			2	9
1495	4	0¾	1	7½	2	11	3	0	2	4½			2	4
1496	5	5½	1	9	3	2			3	0			2	10
1497	5	1	2	0	3	7½			2	9½			3	2
1498	5	5½	2	2¼	4	5			4	6½	4	9	3	9
1499	4	9	1	10¼	3	5¾			3	7¼			3	0
1500	6	1½	1	11	3	8¼	4	4½	3	6½	3	9¼		
1501	8	5¼	2	1½	3	0½	4	8	4	6			4	0
1502	8	0¼	2	4½	4	0¼	6	8	3	7¾			2	5½
1503	6	3¾	1	11¾	4	0¼	5	3	3	5			3	8
1504	5	0¼	2	4	5	0	4	0¼	3	10				
1505														
1506	5	4¾	1	10¼	2	10½			3	3¼	4	0	3	4
1507	5	6½	2	1	2	9¾			2	10	5	4	3	3½
1508	3	10½	2	1	2	10			3	2¼	2	7¼	2	7¾
1509	3	0	1	7¾	3	2			3	1¼	3	11	2	5¾
1510	4	0	1	10¼	4	3	3	5¼	2	5¼			5	3½
1511	5	8½	1	5¼	3	1¼	5	4	3	3½			4	0
1512	9	1¼							4	3¾				

	Wheat (per qtr)		Oats (per qtr)		Barley (per qtr)		Rye (per qtr)		Malt (per qtr)		Peas (per qtr)		Beans (per qtr)	
	s	d	s	d	s	d	s	d	s	d	s	d	s	d
1513	6	0½			5	0					6	1		
1514	5	4	1	8	5	10	4	8	2	11½	7	6		
1515	6	9¾	2	3	2	10	2	8	3	7¼			4	0
1516	5	3½	1	10	3	4½	3	7½	3	10¼	5	2		
1517	6	5			3	4	2	8	3	11¼				
1518	5	11½	2	2¾	3	6	4	0	5	6¾			4	4½
1519	7	2	2	3¾	3	11¾	3	6½	4	8			7	4
1520	9	4½	3	4	5	2	4	11	6	5½				
1521	7	8½	2	8	5	9½			5	7				
1522	6	0¼	2	1½	3	11			4	3½			3	8
1523	5	6	2	3	3	0			3	4				
1524	5	1½	2	10¼	4	0			4	8½			3	1¼
1525	5	5	2	4¾	4	3	6	10	4	8¼			3	10
1526	6	2½	2	10	6	6			5	11¼	7	6	6	8
1527	12	11	3	9¼	5	7¾	12	0¾	9	2½			7	0
1528	8	10¼	3	0	6	7	7	5	5	8			5	10½
1529	8	10	2	4½	2	4½			5	10½	7	1½	5	4
1530	8	5	2	10½	5	0½	5	10	5	5			4	4
1531	8	2¼	3	4½	7	4			8	1				
1532	8	0	3	2¼	5	5½			6	10¼	9	4	6	4¼
1533	7	8	2	9½	4	1¾			5	5¼	4	7	5	4
1534	7	0	3	8	4	0	16	0	3	9			8	0
1535	10	3½	3	4¾	4	5	6	6½	6	0			8	0
1536	10	7¼	3	10¾	4	1¼	8	0	5	5			3	9
1537	7	1	2	8½	5	0	5	4	3	8¾			3	11
1538	6	11½	2	9½	4	7½	5	5	5	0			6	8
1539	5	7¼	2	8	5	4			5	2			5	0
1540	5	8½	3	0½	5	4							5	4
1541	9	0¼	2	10½	4	6			4	10	5	0		
1542	7	11¼	3	0	6	4			4	4¾			5	8
1543	9	3¼	3	0					4	8				
1544	9	0¼	3	4										
1545	15	6¾	4	8	9	0			10	0				
1546	8	3½	4	2	4	0			10	8				
1547	4	11	3	1	3	4			5	4	5	0	6	0
1548	8	1¾	3	6½	3	11½			6	10				
1549	16	4	6	0	11	4			9	4				
1550	18	0	6	8					10	8				
1551	20	4	4	0					5	4				
1552	10	6¾	6	8	8	0			8	0			12	8
1553	10	0	5	4	10	0			9	4				
1554	18	8¼	6	0							6	8	6	8
1555	22	0½	6	0	21	4					16	6¼	18	0

	Wheat (per qtr)		Oats (per qtr)		Barley (per qtr)		Rye (per qtr)		Malt (per qtr)		Peas (per qtr)		Beans (per qtr)	
	s	d	s	d	s	d	s	d	s	d	s	d	s	d
1556	28	5½			15	5			24	0	16	7	18	3½
1557	8	4¾	5	8	6	6					9	4		
1558	9	3½	5	5½	11	4			11	4			10	0
1559	11	0¾	5	7¾	6	8							9	0
1560	14	2¾	6	6	9	0			9	6	6	8		
1561	15	8	6	2½	6	10½			8	0			13	0½
1562	10	11¼	6	4¾	8	5¼			13	0			16	0
1563	19	9¾	7	0	11	6¼			10	8	9	4	16	0
1564	10	10½	5	11½	7	0			13	4	7	10½		
1565	10	7	6	9½	8	1					9	3½	10	11¼
1566	16	5¼	6	4½	7	10	13	4	11	8			10	0
1567	11	1	5	10¾	10	9½			10	0	12	2¼		
1568	11	3½	6	6¼	11	0					9	9		
1569	11	9¼			6	2½			9	8	8	7½		
1570	9	10	5	5					7	2½	8	8		
1571	12	5½	5	7	8	0			8	0	7	4½		
1572	13	6¾	5	5¼	9	4	12	0	8	4	10	5½		
1573	26	3¾	7	0	11	8	18	0	18	0			10	10
1574	14	2¾	5	8½	8	9¼			10	0	12	1½	15	0
1575	15	11	6	5					10	10¾	10	9½		
1576	22	2½	5	6½					14	7½	11	0¼	13	1¼
1577	20	2	6	8					15	0	10	5¼	11	0
1578	17	4¼	4	3¼	8	0	17	8	13	0	10	0	9	0
1579	17	6¼	6	7¾			19	6½	12	5	11	0¼		
1580	20	0	5	0	14	2½	18	2¾	14	4½			12	1
1581	21	5¼	6	4¼			20	0	13	2½	10	8	12	8
1582	19	1½			13	0	22	0	12	3½	12	0	12	1½
1583														
1584														
1585	9	0			4	0								
1586	7	6												
1587	64	0												
1588	14	2½												
1589														
1590														
1590														
1591														
1592														
1593														
1594	56	0												
1595	53	0												
1596	80	0	31	0										
1597	92	0												

	Wheat (per qtr)		Barley (per qtr)		Rye (per qtr)		Malt (per qtr)		Peas (per qtr)		Beans (per qtr)	
	s	d	s	d	s	d	s	d	s	d	s	d
1598	56	8										
1599	39	2										
1600	36	8										
1601	34	10										
1602	29	4										
1603	35	4										
1604	30	8	14	0	15	0			15	0	15	0
1605	35	10										
1606	33	10										
1607	36	8										
1608	56	8										
1609	50	0										
1610	35	10										
1611	38	8										
1612	42	4										
1613	48	8										
1614	41	8½										
1615	38	8										
1616	40	4										
1617	48	8										
1618	46	8										
1619	35	4										
1620	30	4										
1621	30	4										
1622	58	8										
1623	52	0	12	0	20	0	12	0	16	0	16	0
1624	48	0										
1625	52	0										
1626	49	4										
1627	36	0										
1628	28	0										
1629	42	0										
1630	55	8										
1631	68	0										
1632	53	4										
1633	58	0										
1634	56	0										
1635	56	0										
1636	56	8										
1637	53	0										
1638	57	0										
1639	44	10										

	Wheat (per qtr)		Oats (per qtr)		Barley (per qtr)		Malt (per qtr)		Beans (per qtr)		Cattle	
	s	d	s	d	s	d	s	d	s	d	s	d
1640	44	8										
1641	48	0										
1642	60	2										
1643	59	10										
1644	61	3									£5.0.0.	
1645	51	3										
1646	48	0										
1647	73	8										
1648	85	0										
1649	80	0										
1650	76	8										
1651	73	4										
1652	59	6										
1653	31	6										
1654	23	1										
1655	29	7										
1656	38	2										
1657	41	5										
1658	57	9										
1659	58	8										
1660	50	2	16	0	20	0	20	0	24	0		
1661	62	2										
1662	65	9										
1663	50	8										
1664	36	0										
1665	43	0										
1666	32	0										
1667	32	0										
1668	35	6										
1669	39	0										
1670	37	0										
1671	37	4										
1672	36	5										
1673	41	5										
1674	61	0										
1675	57	5										
1676	33	9										
1677	37	4										
1678	52	5										
1679	53	4										
1680	40	0										
1681	41	5										

	Wheat (per qtr)			Wheat (per qtr)			Wheat (per qtr)	
	s	d		s	d		s	d
1682	39	1	1706	26	0	1730	36	6
1683	35	6	1707	28	6	1731	32	10
1684	39	1	1708	41	6	1732	26	8
1685	41	5	1709	78	6	1733	28	4
1686	30	2	1710	78	0	1734	38	10
1687	22	4	1711	54	0	1735	43	0
1688	40	10	1712	46	4	1736	40	4
1689	26	8	1713	51	0	1737	38	0
1690	30	9	1714	50	4	1738	35	6
1691	30	2	1715	43	0	1739	38	0
1692	41	5	1716	48	0	1740	50	8
1693	60	1	1717	45	8	1741	46	8
1694	56	10	1718	38	10	1742	34	0
1695	47	1	1719	35	0	1743	24	10
1696	63	1	1720	37	0	1744	24	10
1697	53	4	1721	37	6	1745	27	6
1698	60	9	1722	36	0	1746	39	0
1699	56	10	1723	34	8	1747	34	10
1700	35	6	1724	37	0	1748	37	0
1701	33	5	1725	48	6	1749	37	0
1702	26	2	1726	46	0	1750	32	6
1703	32	0	1727	42	0	1751	38	6
1704	41	4	1728	54	6			
1705	30	0	1729	46	10			

	Wheat (per qtr)		Butter (per lb)		Cheese (per lb)		Meat (per lb)		Hens		Bacon (per lb)	
	s	d	s	d	s	d	s	d	s	d	s	d
1752	41	10	3½d–4d		2½d–3d		2½d–3d					
1753	44	8		6				2¾				6
1754	34	8										
1755	33	10										
1756	45	3										
1757	53	4										
1758	50	0										
1759	39	10										
1760	36	6										
1761	30	3										
1762	39	0										
1763	40	9										

	Wheat (per qtr)		Barley (per qtr)		Oats (per qtr)		Butter (per lb)		Cheese (per lb)		Meat (per lb)	
	s	d	s	d	s	d	s	d	s	d	s	d
1764	46	9						$6\frac{1}{4}$		3		3
1765	48	0						8		$3\frac{1}{2}$		$3\frac{1}{2}$
1766	43	1										
1767	47	4										
1768	53	9										
1769	40	7										
1770	43	6										$2\frac{1}{2}$
1771	47	2										
1772	50	8										
1773	51	0										
1774	52	8										
1775	48	4										
1776	38	2										
1777	45	6										
1778	42	0	20	0	14	6						
1779	33	8	26	0	13	6						
1780	35	8										
1781	44	8										
1782	47	10										
1783	52	8										
1784	48	10										
1785	51	10										
1786	38	10										
1787	41	2										
1788	45	0										
1789	51	2										
1790	54	9	26	3	19	5				4		$3\frac{3}{4}$
1791	48	7	26	10	18	1				$4\frac{1}{4}$		4
1792	43	0	27	7	16	9		10		4d–5d		4d–$4\frac{1}{2}$d
1793	49	3	31	1	20	6				$4\frac{3}{4}$		$4\frac{1}{4}$
1794	52	3	31	9	21	3				5		$4\frac{1}{2}$
1795	75	2	37	5	24	5				$5\frac{1}{4}$		$4\frac{1}{2}$
1796	78	7	35	4	21	10				$5\frac{1}{2}$		6
1797	53	9	27	2	16	3				$5\frac{1}{2}$		$6\frac{1}{2}$
1798	51	10	29	0	19	5				$5\frac{3}{4}$		$5\frac{1}{2}$
1799	69	0	36	2	27	6				6		$5\frac{1}{2}$
1800	113	10	59	10	39	4				$6\frac{1}{4}$		$6\frac{3}{4}$
1801	119	6	68	6	37	0				$6\frac{1}{2}$		8
1802	69	10	33	4	20	4				$6\frac{1}{2}$		$8\frac{1}{4}$
1803	58	10	25	4	21	6				$6\frac{3}{4}$		$7\frac{1}{2}$
1804	62	3	31	0	24	3				7		8
1805	89	9	44	6	28	4				$7\frac{1}{4}$		$7\frac{1}{4}$

	Wheat (per qtr)		Barley (per qtr)		Oats (per qtr)		Cheese (per lb)		Meat (per lb)		Wool (per lb)		Sheep	
	s	d	s	d	s	d	s	d	s	d	s	d	s	d
1806	79	1	38	8	27	7						7¼		7½
1807	75	4	39	4	28	4						7½		7
1808	81	4	43	4	33	4						7¾		6¾
1809	97	4	47	0	31	5						8		6¾
1810	106	5	48	1	28	7						8		8½
1811	95	3	42	3	27	7						8¼		8¾
1812	126	6	66	9	44	6						8½		8¼
1813	109	9	58	6	38	6		8¾		8¾				
1814	74	4	37	4	25	8		8¾		9½			38	0
1815	65	7	30	3	23	7		8		8¼				
1816	78	6	33	11	27	2		6½		7¼				
1817	96	11	49	4	38	5		5½		6				
1818	86	3	53	10	33	5		6		7¼				
1819	74	6	45	9	28	2		8		7¾				
1820	67	10	33	10	24	2		7		7½			35	0
1821	86	1	26	0	19	6		5½		6¾				
1822	44	7	21	10	18	1		4¼		5	1	6	45	0
1823	53	4	31	6	22	11		4		4¾		18		
1824	63	11	36	4	24	10		4		5		18		
1825	68	6	40	0	25	8		5½		6½	1	0	33	6
1826	58	8	34	4	26	8		6½		6¼	1	0		
1827	58	6	37	7	28	2		5¼		6½		10		
1828	60	5	32	10	22	6		6		5¾		9		
1829	66	3	32	6	22	9		6		5½		8		
1830	64	3	32	7	24	5		5¾		4¾	1	0	20	0
1831	66	4	38	0	25	4		5¾		5¾		14	23	0
1832	58	8	33	1	20	5		5¾		5¼	1	0	27	0
1833	52	11	27	6	18	5		6		5½		18	27	0
1834	46	2	29	0	20	11		6½		5		19	27	0
1835	39	4	29	11	22	0		6¼		4¾		18	20	0
1836	48	6	32	10	23	1		6½		5½		20¼	23	6
1837	55	10	30	4	23	1		7		5¾		16	27	0
1838	64	7	31	5	22	5		6¾		5½		17¼	30	0
1839	70	8	39	6	25	11		6½		5¾		18	29	0
1840	66	4	36	5	25	8							23	9
1841	64	4	32	10	22	5					1	0	29	0
1842	57	3	27	6	19	3						13	21	0
1843	50	1	29	6	18	4						10	26	6
1844	51	3	33	8	27	7						13	26	0
1845	50	10	31	8	22	6						14¼	31	0
1846	54	8	32	8	23	8						12½	40	0
1847	69	9	44	2	28	8						13	33	0

	Wheat (per qtr)		Barley (per qtr)		Oats (per qtr)		Meat (per lb)		Wool (per lb)		Sheep	
	s	d	s	d	s	d	s	d	s	d	s	d
1848	50	6	31	6	20	6				9	27	6
1849	44	3	27	9	17	6				11	25	0
1850	40	3	23	6	16	5			1	0	27	0
1851	38	6	24	9	18	7			1	0	25	6
1852	40	9	28	6	19	1				13¼	30	0
1853	53	3	33	2	21	0				17½	32	0
1854	72	5	36	0	27	11				13	30	0
1855	74	8	34	9	27	5				15	28	0
1856	69	2	41	1	25	2				16½	32	3
1857	56	4	42	1	25	0		6¾		17	37	0
1858	44	2	34	8	24	6		6¾		15½	30	0
	Wheat (per cwt)		**Barley (per cwt)**		**Oats (per cwt)**							
	s	d	s	d	s	d						
1859	10	3	9	5	8	4				18¾		
1860	12	5	10	3	8	9		7¼		19½	34	0
1861	12	11	10	1	8	6		6¾		17	33	0
1862	12	11	9	10	8	1		6½		18¼	37	0
1863	10	5	9	6	7	7		7¼		21	40	0
1864	9	4	8	5	7	3		7¼		25½		
1865	9	9	8	4	7	10		8		21¾	48	0
1866	11	8	10	6	8	10				19	50	6
1867	15	0	11	2	9	4		7¾			44	0
1868	14	11	12	0	10	1		7¼		16	26	0
1869	11	3	11	0	9	4		6½		13½	35	0
1870	10	11	9	8	8	2		8		13		
1871	13	3	10	2	9	0		8½		17½	45	0
1872	13	4	10	5	8	4		8¾		21½		
1873	13	8	11	4	9	1		9½		19	50	0
1874	13	0	12	7	10	4		9		18	41	0
1875	10	6	10	9	10	3		9½		18½		
1876	10	9	9	10	9	5						
1877	13	3	11	1	9	4		10		17½	58	6
1878	10	10	11	3	8	9		9½		15½		
1879	10	3	9	6	7	10		9¼	1	0	43	0
1880	10	4	9	3	8	3				16	53	0
1881	10	7	8	11	7	10					54	0
1882	10	6	8	9	7	10					60	0
1883	9	8	8	11	7	8						
1884	8	4	8	7	7	3						

	Wheat (per cwt)		Barley (per cwt)		Oats (per cwt)		Cattle (per live cwt)	
	s	d	s	d	s	d	s	d
1885	7	8	8	5	7	5		
1886	7	3	7	5	6	10		
1887	7	7	7	1	5	10		
1888	7	5	7	10	6	0		
1889	6	11	7	3	6	4		
1890	7	5	8	0	6	8		
1891	8	8	7	11	7	2		
1892	7	1	7	4	7	1		
1893	6	2	7	2	6	9		
1894	5	4	6	10	6	2		
1895	5	5	6	2	5	2		
1896	6	1	6	5	5	4		
1897	7	0	6	7	6	1		
1898	7	11	7	7	6	7		
1899	6	0	7	2	6	1		
1900	6	3	7	0	6	4		
1901	6	3	7	1	6	7		
1902	6	7	7	2	7	3		
1903	6	3	6	4	6	2		
1904	6	7	6	3	5	10		
1905	6	11	6	10	6	3		
1906	6	7	6	9	6	7		
1907	7	2	7	0	6	9		
1908	7	6	7	3	6	0		
1909	8	7	7	6	6	9		
1910	7	5	6	6	6	3		
1911	7	5	7	8	6	9		
1912	8	1	8	7	7	9		
1913	7	5	7	8	7	9		
1914	8	2	7	7	7	6		
1915	12	4	10	5	10	10		
1916	13	8	15	0	12	0		
1917	17	8	18	1	17	11		
1918	17	0	16	6	17	9		
1919	17	0	21	2	18	9		
1920	18	10	25	0	20	5		
1921	16	8	14	7	12	3		
1922	11	2	11	2	10	5		
1923	9	10	9	5	9	7		
1924	11	6	13	1	9	9	60	0
1925	12	2	11	9	9	9	59	11
1926	12	5	10	4	9	0	55	10

	Wheat (per cwt)		Barley (per cwt)		Oats (per cwt)		Cattle (per live cwt)		Potatoes	
	s	d	s	d	s	d	s	d	s	d
1927	11	6	11	9	9	1	50	6		
1928	10	0	11	0	10	5	55	0		
1929	9	10	9	11	8	10	52	4		
1930	8	0	7	11	6	2	52	1		
1931	5	9	7	11	6	3	47	6		
1932	5	11	7	7	7	0	44	10		
1933	5	4	7	11	5	7	39	10		
1934	4	10	8	8	6	3	39	2		
1935	5	2	7	11	6	8	36	1		
1936	7	2	8	3	6	4	37	6		
1937	9	4	10	11	8	7	42	2		
1938	6	9	10	2	7	7	42	10		
1939	5	0	8	10	6	11	43	10		
1940	10	0	18	2	13	4	60	0		
1941	14	8	24	0	14	8	62	7		
1942	15	11	45	8	14	11	67	8		
1943	16	3	31	5	15	8	69	3		
1944	14	11	26	5	16	3	70	9		
1945	14	5	24	5	16	5	72	10		
1946	14	10	24	3	16	3	77	0		
1947	16	9	24	0	18	3	89	11		
1948	21	0	26	10	20	10	98	8		
1949	23	3	25	10	21	0	103	6		
1950	25	10	27	11	21	7	105	3		
1951	28	8	38	10	26	2	114	0		
1952	29	7	32	7	26	9	125	9		
1953	31	2	30	1	24	7	132	7		
1954	28	3	25	9	22	7	134	7		
1955	22	11	26	0	26	3	154	1		
1956	25	6	25	8	24	8	117	6		
1957	21	7	23	2	22	10	128	7		
1958	21	9	22	11	23	10	146	8		
1959	21	0	22	7	22	7	158	0		
1960	21	4	21	3	22	6	149	9		
1961	20	7	19	10	19	5	125	4		
1962	21	10	23	0	23	0	143	2		
1963	20	11	20	8	21	0	134	10		
1964	21	11	21	2	21	5	163	3		
1965	22	6	22	10	22	8	174	9		
1966	22	4	22	1	22	8	167	8		
1967	25	2	24	7	27	0			282	0

	Wheat (millable) (per cwt)		Barley (per cwt)		Oats (per cwt)		Sugar Beet (per ton at 16% sugar)		Potatoes (per ton)	
	s	d	s	d	s	d	s	d	s	d
1968	27	5	25	2	27	10	136	6	297	6
1969	29	0	26	0	27	10	136	6	302	6
1970	30	3	27	0	27	10	136	6	317	6
1971	32	7	29	0	28	10	152	0	331	0
(new decimalisation)										
	(£1·63)		(£1·45)		(£1·44)		(£7·60)		(£16·55)	
	£		£		£		£		£	
1972	1·72		1·56		1·51		8·00		16·55	
	(per tonne)		(per tonne)		(per tonne)					
1973	36·70		33·20		32·00		8·00		17·00	
1974	39·62		35·73		34·20		9·41		22·00	
1975	51·80		46·80		44·60		13·96		28·00	
1976	65·00		63·00		58·00		18·41		91·44	
1977	85·86		84·33		77·73		19·89		142·82	

Cattle (Fat) (per cwt)	Fat Lambs and Sheep (per lb)	Fat Pigs (per score)	Milk (per gallon)	Wool (per lb)
s d	s d	s d	s d	s d
200 0	3 6·25	47 11	3 8·86	4 5·25
215 0	3 7·75	48 5	3 9·26	4 5·25
222 6	3 10·75	50 11	3 11	4 5·25
247 0	4 5·5	58 7	4 5	4 6·5
(£12·35)	(22·3p)	(£2·93)	(22·1p)	(22·7p)
£	p	£	p	p
13·20	24·03	2·81	23·01	23·00
13·20	26·05	3·46	24·61	25·00
—	29·05	3·49	26·27	34·75
—	35·05	—	34·75	31·00
23·52	37·06	5·86	37·09	31·00
29·51	50·00	6·33	44·00	37·97